Improve
FUEL AVERAGE
(MPG / KMPL)

by

PRAVIN SHINDE

Preface

When we start dealing with any new problem, many times it is observed that initially it is difficult to understand the problem in its true perspective. When we go on observing the problem very carefully, all of a sudden you will find that you have reached to root cause of the problem & then every thing is very simple.

In case of fuel average also, since I was dealing with City operations I used to find large variations & could not relate the reasons for these variations even after applying all my knowledge gathered both in engineering college & subsequent reading of various automobile books. But I continued my efforts to analyze fuel average.

When I started checking fuel consumption of engine under no load condition (i.e. vehicle stationary & only engine running) at different r.p.m.s, I was shocked to see that fuel consumption/hr at maximum engine r.p.m. under no load condition was more than average fuel consumption/hr for operating the vehicle (calculated from average consumption & operating hours). Initially it was hard to believe. But the fact was confirmed by making observations on different vehicles also. On detailed observations it was found that average 'no load fuel consumption' of the engine in the normal engine r.p.m. range was almost 70 % of the total average consumption/hr for operating the vehicle. In other words 70 % of fuel consumption was due to engine friction only. This made me realize the exact meaning of a sentence in our text books – 'Under part load conditions engine efficiency is low.'

Based on the above observations then I developed a method of accounting 'no load fuel consumption' of engine as fixed part & remaining part as variable. This method of accounting was then tested under different conditions, on varieties of vehicle models & found to be tallying with theoretical calculations.

With this, basic problem of understanding all variations in different conditions was completely solved. Details of all these points are given in the book. Moreover, this method enables to prepare a mathematical model for all variations in different conditions & graphical presentation of the same gives a clear picture at a glance about how fuel average varies under different conditions.

But the main problem was educating the drivers on mass scale. After all if we have to achieve higher mileage for the fleet, drivers are the key persons who are going to implement it.

To educate drivers on mass scale, I had developed very simple system to practically demonstrate variations in fuel consumptions under different conditions just in few seconds (for each point) & that resulted in very good response from drivers.

Details of all the methods both for physical fuel measurement in just few seconds & also by just carrying out natural retardation test (this also in few seconds & without any instruments) are given in the book.

Also tips for driving for better fuel economy & also various points for design improvements based on the observations arising out of this analysis are also given.

During my long 36 yr.s of experience, I have always observed that when we start working on any problem, initially progress is very slow. But once we get acquainted with basic things to make our concepts clear, then we can go very fast in solving even any complex problems. I am quite sure that once you get acquainted with this, you will go miles ahead.

So, best of luck!

PRAVIN SHINDE

Table of Contents

1.Introduction

Every body is aware that limited reserves of crude petroleum on one hand & increasing demand on the other hand is resulting in increase in fuel prices at an alarming rate.

Therefore everybody is conscious about saving of fuel. At present, even car manufacturing companies are advertising about fuel average (MPG or KMPL) of their car models. This is actually fuel average that we can get when we are operating the vehicle in top gear at certain speed, say 60 kms/hr. However, when we are operating the vehicle in different conditions there is wide variation in fuel consumption.

For example –

i. if we operate the vehicle at a speed above 120 kms/hr instead of 60 kms/hr, then increase in fuel consumption is more than 100 % & at still higher speeds, % increase rises much more sharply.

ii. When we are operating the vehicle in lower gears, increase in fuel consumption can be more than 100 %.

In short, fuel consumption is changing continuously as per the operating condition. What trip average we are getting is the average of all these values.

Therefore, we need to understand how fuel consumption varies under different conditions, so that we can avoid or minimize operation in high fuel consumption zone.

To have proper visualization, it is necessary to have proper accounting system. Good accounting system helps us to understand how expenditure is made under different heads, so that clear picture will be in front of our eyes. Our aim is not to prepare a statement of accounts, but prepare simple accounting system so that entire scenario we can visualize without any difficulty.

But before proceeding to accounting system, let us first understand few important factors that pose difficulties / create confusions in understanding variations in fuel consumption.

1.1 Variation in engine efficiency

1.1 **Variation in engine efficiency** – Engine efficiency varies to a great extent depending on the load on the engine. Under part load conditions, engine efficiency is low. This variation in engine efficiency makes it difficult to have proper accounting of fuel consumption.

To overcome this problem fuel consumption is divided into fixed part & variable part as we do in case of working out manufacturing cost.

1.1.A 'No load fuel consumption' of engine is fixed part of fuel consumption & needs to be accounted separately –

When engine is running at any r.p.m. (**R**evolutions **P**er **M**inute) under 'no load condition' (i.e. when vehicle is stationary & only engine is running), it consumes some amount of fuel. In this case since there is no output taken from engine, it is clear that this fuel consumption is utilized to overcome internal friction of the engine. This fuel consumption is fixed part of fuel consumption due to engine friction.

Once we separate out fixed part of fuel consumption, remaining part is obviously proportional to load on the engine. This gives us flexibility of accounting energy consumption & then work out its fuel equivalent.

This point is explained in detail in chapter on 'Engine Friction'.

1.2 - For Proper Accounting, only Actual Energy Consumed should be accounted–

In automobiles we many times come across the situation where fuel is consumed & energy is produced. But this energy is not consumed. In such case there is no point in accounting for fuel consumed. This energy is then subsequently consumed & should be accounted at that time only.

Following examples will make the point clear –

1.2.A – Acceleration – During acceleration as vehicle speed increases kinetic energy of the vehicle is increased. For providing this increase in kinetic energy, extra fuel is required. Therefore during acceleration there is extra fuel consumed, in addition to fuel required for normal steady speed running.

But this energy is not consumed. It is stored in the vehicle in the form of kinetic energy which can be utilized for running extra distance without consuming any fuel at a later stage.Therefore fuel average for the trip remains unaffected & hence no need to account for extra fuel consumed during acceleration.

1.2.B – Moving on a gradient – In this case also extra fuel is required to provide for increase in 'Potential Energy' of the vehicle, but energy is not consumed. It is stored in the vehicle.

Therefore in this case also there is no need to account for extra fuel consumed.

1.2.C – When Vehicle Slows Down, there is no fuel consumption, but kinetic energy of the vehicle is consumed –

When we allow the vehicle to slow down just by releasing accelerator pedal, then fuel supply to the engine is cut off & vehicle is running using only kinetic energy of the vehicle.

In this case, there is no fuel consumption, but energy is consumed. This energy consumption is same as in normal course for running the vehicle at that speed & gear.

Actually, this energy is produced at the cost of extra fuel consumed during acceleration. In this case, even though there is no fuel consumption, energy consumption needs to be accounted for. This energy consumption then can be converted to fuel consumption.

1.2.D – Braking Losses – During braking, there is no fuel consumption, but there is consumption of energy. Therefore, it is necessary to account for this energy consumption.

From the above examples it can be seen that if we decide to account only fuel consumed, it will only lead to confusion. Moreover loss due to braking cannot be accounted for. Therefore for proper understanding of how fuel is utilized, we need to account for actual energy consumption.

As given in section 1.1 we have separated out fixed part of fuel consumption & therefore remaining part of fuel consumption is directly proportional to energy consumption. Therefore it is possible to convert energy consumption to equivalent fuel consumption.

Now a days fuel consumption is being expressed in **M**iles **P**er **G**allon i.e. **MPG** or **KM**s **P**er **L**iter i.e. **KMPL**. In olden days it was being expressed in gallons/100 miles or liters/100 kms.

But we will be expressing fuel consumption in cc/km for the sake of convenience. (1 liter = 1000 cc)

Expressing fuel consumption in cc/km has the advantage that fuel consumption due to different forces, can be worked out separately & then adding them together we can find total fuel consumption.

1.3 Fuel consumption – We have seen that fuel consumption is directly proportional to energy required to drive the vehicle. Therefore first let us study energy requirement of the vehicle for constant speed running on level road.

1.3.A Energy required to drive the vehicle – When vehicle is in motion, there are forces acting on it to oppose the motion. Therefore engine has to supply this energy required to overcome forces acting on it.

At constant speed running, on level road, only forces acting on the vehicle are rolling resistance, air resistance & engine friction.

i. Rolling Resistance - This is due to rolling resistance of tyre & depends on total weight of the vehicle. This force does not vary with vehicle speed or gear in which vehicle is operated. Therefore it is constant for any given vehicle with its loading condition.

ii. Air Resistance – This force is proportional to square of vehicle speed & therefore goes on increasing rapidly with increase in vehicle speed.

iii. Engine Friction – We will see about this in detail in the chapter on 'engine friction'. Force or fuel consumption due to engine friction depends upon vehicle speed & also gear in which vehicle is operated. (actually it depends upon engine r.p.m.)

Sum total of these forces gives us total force acting on the vehicle. i.e. –

$$F = F_{rr} + F_{ar} + F_{ef}$$

In the above equation suffixes rr, ar & ef are added to force to indicate force due to rolling resistance, air resistance & engine friction. Same convention will be followed for fuel consumption – FC.

(We will study force & fuel consumption due to each of the factors in detail in respective chapters.)

We know that energy is given by the equation –

Energy E = Force x distance

Therefore –

$$E/km = F \times 1 = F ....eq^n \ (1.1)$$

In the above equation force is in kg & is noted by 'F' & distance is 1 km. Therefore energy E is in kg-km.

We will be expressing energy E in kg-km instead of normal practice of expressing it in kg-m for the sake of convenience.

By using eq^n (1.1) we can find out energy required for overcoming different forces separately or we can add forces together to work out total energy required.

1.3.B – Fuel Consumption for Constant Speed Running on Level Road –

As seen above, when we account for fixed part of fuel consumption separately, then remaining part is directly proportional to energy requirement. Let constant of proportionality be noted by Cf – conversion factor. Then –

Fuel Consumption –
$$FC = Cf \times E \ ..eq^n \ (1.2)$$
where FC – fuel consumption in cc/km
Cf – conversion factor in cc/kg-km
& E - energy in kg-km/km

From eq^n (1.1) we know that E = F, therefore above equation reduces to –

$$FC = Cf \times F \ ... eq^n \ (1.3)$$

From the above equation we can see that once we know the value of any force acting on the vehicle, we can easily calculate fuel consumption in cc/km by simply multiplying force F by Cf. Also from fuel consumption we can find out value of force.

Sum total of fuel consumption due to rolling resistance + air resistance + engine friction will give us total fuel consumption FC for constant speed running, on level road. i.e. –

$$FC = FCrr + FCar + FCef \ \ ... \ eq^n \ (1.4)$$

Above equation (1.4) can also be written as –

$$FC = Cf \ x \ (Frr + Far + Fef)$$

Since fuel consumption FC is expressed in cc/km, from the value of FC we can easily calculate KMPL as given below –

$$KMPL = 1000 \ / \ FC \ ... \ eq^n \ (1.5)$$
$$(since \ 1 \ liter = 1000 \ cc)$$

Above KMPL figure can be converted to MPG by just multiplying it with 2.35. i.e. -

$$MPG = 2.35 \ X \ KMPL \ ... \ eq^n \ (1.6)$$

Thus by following the method of accounting engine friction separately, fuel consumption is directly proportional to force & this makes things much easier to understand.

Advantages of this method –

1. Fuel Consumption can be calculated easily using simple mathematical formula for any vehicle speed in any gear.

2. Fuel consumption due to different factors can be worked out separately which gives clear idea about relative effect of each factor.

3. Since fuel consumption is worked out from force required to drive the vehicle & then just multiplied by conversion factor – Cf, this method can be easily used for any type of fuel used whether Petrol, CNG, LPG, Diesel or any other type of vehicle. Only thing that will vary is conversion factor which we can find out practically. Moreover, for achieving fuel economy, we are interested in knowing % variations in fuel consumption in different conditions of vehicle speed & gear, therefore we can simply compare total force acting on the vehicle under different conditions to understand variations in fuel consumption.

4. Practical Testing – Since fuel consumption is directly proportional to force, it becomes very easy to find out fuel consumption in different gears at different vehicle speeds by practically measuring force.

Total force acting on the vehicle can be measured by 'natural retardation test' i.e. simply allowing vehicle to slow down & measuring time in seconds. This testing does not require any instruments & can be done in just few seconds.

Moreover, it is possible to check forces due to rolling resistance, air resistance & engine friction practically by natural retardation tests.

Further <u>it is possible to directly compare fuel average (MPG / KMPL) for any two conditions of vehicle speed & gear by above method in just few minutes, that too without any instruments, without any calculations & need absolutely no data of the vehicle</u>.

We will see in detail about these tests in the chapter on 'Practical Testing'.

It is important for us to know how fuel consumption due to each of the above mentioned three factors vary, particularly, quantitative effect of each factor in relation to other factors.

We have already seen that fuel consumption due to rolling resistance is constant & that due to air resistance is proportional to V^2.

In the chapter on 'Engine Friction' we will see how fuel consumption due to engine friction i.e. FCef varies with vehicle speed & gear.

For better understanding of the quantitative effect of each of the factor, we have taken a model car with following specification for analysis –

Specifications of Model Car

W	1600 kg	Gear ratios	
A	2.35 sq.m.	1st	3.91
Engine	1200 cc	2nd	2.24
Tyre dia.	0.58 meters	3rd	1.44
Diff. ratio	3.56	4th	1
Co-efficients		top	0.77
Crr	0.01		
Cd	0.45		
Cf	0.8		

W – total weight of the vehicle

A- frontal area of the vehicle

Variations in fuel consumption for entire speed range, in different gears calculated from basic formula are shown in the chapter on 'Variations in Fuel Consumption' Also various graphical presentations are given for better visualization. Graph of fuel average for different vehicle speeds & for different gears shown on the cover page is the actual graph plotted from calculated values using above method for model car.

Even though as an illustration, sample analysis is done for model car as above, formula given are general & can be used for any type of vehicles i.e. right from scooters, motor cycles, 3 wheelers, cars, buses, trucks or tractor trailers etc. & also for any type of fuel i.e. petrol, CNG, LPG or Diesel.

Once we understand variations in fuel consumption for any one type of vehicle, we can very easily understand variations for any other types of vehicles because basic pattern will remain the same.

In this section we have seen how fuel consumption varies for constant speed running in different gears.

In section 1.2 we have seen that there is no need to account for extra fuel consumed for acceleration or moving up a gradient.

Therefore only factors that need to be accounted for are braking losses & idling losses. We will see these in next section.

1.4 <u>Other Losses</u> - In addition to above points we need to consider the following points. These were deliberately not included in the points given above, as they are of different nature altogether as explained below –

1.4.A - Braking losses – During braking, accelerator pedal is in released position & as engine speed is high, fuel supply to the engine is cut off. Therefore during the period of braking, there is no fuel consumption, but vehicle is running using kinetic energy of the vehicle.

Even though during braking there is no fuel consumption, part of kinetic energy of the vehicle is consumed. This kinetic energy we had gained at the cost of extra fuel consumed during acceleration.

Therefore this energy consumption needs to be accounted for. We will study this in detail in the chapter on braking losses. We will see how % increase in fuel consumption due to braking losses can vary from less than 5 % to more than 40 % & how to reduce the losses.

1.4.B - Engine Idling losses – At signals or in traffic jams, we come across the situation when vehicle is stationary, but engine is running. Under this condition there is fuel consumption but distance covered is zero. We will see in the chapter on 'Idling losses' that % increase in fuel consumption due to idling losses can be around 5 % in City operations when % idling time to total time is as high as 25 %.

In above two cases – 1.4.A & 1.4.B, we cannot express fuel consumption/km as done in case of other items listed above. Moreover they are not of continuous nature. Therefore they need to be accounted separately.

We will see in detail how to account for the above factors in respective chapters.

Therefore accounting fuel for 3 basic forces i.e. Frr, Far & Fef & then adding losses for braking & idling we get the completer picture about average fuel consumption for the trip.

After all our aim is to understand variations in fuel consumption to avoid or minimize the losses so as to get best possible fuel average.

In subsequent chapters, we will study in detail about each of the factor affecting fuel consumption, mentioned above. Graphical presentations of these variations, given in chapter on 'variations in fuel consumption' will help us to give clear picture at a glance & will help us to improve our fuel average.

Driving for better KMPL - Only understanding how fuel average varies is not of any use, unless we prepare our self for action plan. For this, various points that need to be taken care of while driving under practical conditions in City traffic & highway running are discussed in the chapter on 'Driving for Better Fuel Average'.

In this book we have discussed on the parameters that influence fuel consumptions to a large extent due to operating conditions. You can see that these variations are affecting fuel consumption to the tune of 100 % or even more.

But it is true that there are large no. of factors that affect fuel consumption but their effect is comparatively much smaller. Even about engine efficiency, after 5 to 10 years of extensive research can give improvement for 5 to 10 %. In the method followed by us it is taken care of by conversion factor Cf which we are verifying practically.

In conventional method of testing the vehicle, for fuel average for complete trip, even though there is improvement due to some factors, they are being often neglected as the result of such testing are not consistent due to large variations in operating conditions. Methods of practical testing, for predefined conditions only, given in this book, can help in finding out realistic improvement even for small improvement of below 5 %.

---oOo---

2. Rolling Resistance

Rolling resistance is the force required to overcome rolling friction at the wheels. This force is proportional to the load on the wheels & constant of proportionality is called co-efficient of rolling resistance. Therefore force due to rolling resistance – F – is given by the formulae –

$$F = Crr \times W \dots eq^n (2.1)$$

Where, F – force due to rolling resistance
Crr – co-efficient of rolling resistance
& W – total weight of the vehicle

Co-efficient of rolling resistance depends upon construction of tyres & nature of road surface. Tyres are basically of two types – cross ply tyres & radial ply tyres. With radial ply tyres value of co-efficient of rolling resistance is less by nearly 10 % & correspondingly, force due to rolling resistance is also less. It also depends upon the nature of road surface. For concrete roads it is at the minimum, for asphalt road it is more & for rough roads it is still on higher side.

When we are push starting the vehicle, it is in gear & therefore force needed to push the vehicle is more as additional force is required to crank the engine. But when we are pushing the vehicle by keeping it in neutral, it is basically the force required to overcome friction at the wheels. But in this case also in addition to friction at the wheels, there is additional friction at wheel hub bearings & differential. Therefore we will consider the value of co-efficient of rolling resistance including friction in the differential & wheel bearings. Moreover for practical testing to find out the value of co-efficient of rolling resistance vehicle is tested by allowing to slow down in neutral gear as explained in the chapter on 'Practical Testing'. Therefore it is more appropriate to include these frictional forces in differential & wheel hub bearings for calculating rolling resistance.

From equation (1.3) given in chapter 1, fuel consumption FC = Cf x F. Therefore substituting the value of force given in equation (2.1) , we get –

$$FCrr = Cf \times (Crr \times W) \ ... \ eq^n \ (2.2)$$

Now, let us see the value of rolling resistance for our model car. Substituting the values for our model car in equation (2.1) we get –

Rolling Resistance –

$$Frr = 0.01 \times 1600$$

$$= 16 \ kg. \ ... \ eq^n \ (2.1a)$$

Fuel consumption due to rolling resistance in this case will be –

$$FCrr = Cf \times Crr \times W$$

$$= 0.8 \times 0.01 \times 1600$$

$$= 12.8 \ cc/km \ ... eq^n \ (2.2a)$$

Thus we can see that rolling resistance Frr for our model car is 16 kg & fuel consumption due to rolling resistance FCrr is 13 cc/km.

Summary –

It can be seen that both force due to rolling resistance - Frr & fuel consumption due to rolling resistance - FCrr are proportional to vehicle weight W & remain constant. They do not vary with vehicle speed or gear.

---oOo---

3. Air Resistance

When vehicle moves forward at high speed, it pushes the air in front of the vehicle forward, creating an increase in pressure in the front of the vehicle. When the pressure increases, air also starts flowing in other adjacent areas due to pressure difference. At very low speeds there is no appreciable increase in pressure in front of the vehicle. However with increase in vehicle speed rate of rise in pressure is much faster than the rate of flow to the adjoining area, this, results in considerable increase in pressure in front of the vehicle.

To reduce air resistance, front face of the vehicle is kept tilted backward with smooth curvature & also provided with backward curved portion in the side ways direction also. The basic purpose of providing all these curved shapes is to allow smooth flow of air from the front of the vehicle so that pressure build up in front of the vehicle is reduced to minimum. Less pressure build up in front of the vehicle results in less air resistance.

We have seen above, how pressure build up in front of the vehicle causes air resistance. Simultaneously, when the vehicle moves forward with high speed, there is vacuum created at the back of the vehicle. Due to this vacuum, air from surrounding area starts flowing to this area. But with increase in vehicle speed, level of vacuum also increases. Vacuum at back of the vehicle creates a backward thrust on the vehicle. To reduce the formation of vacuum at back of the vehicle, instead of keeping backside face vertical, it is kept sloping & curved for smooth flow of air at the back of the vehicle.

Air Resistance is proportional to the frontal area of the vehicle. However depending upon the aerodynamic design of the body, this force is reduced. Normal practice is to apply a correction factor to the frontal area to get effective frontal area. This is called as 'Coefficient of air drag or air resistance'.

Air Resistance also varies in proportion to the square of vehicle speed – 'v' expressed in m/sec. But our normal practice is to express vehicle speed – V in kms/hr. Therefore for entering value of vehicle speed into formulae we have to first convert value expressed in kms/hr to m/sec. Therefore –

$$v = V \times 1000 / (60 \times 60)$$

$$= V / 3.6 \text{ in m/sec. .. eq}^n \text{ (3.1)}$$

Formulae used to calculate force due to air resistance is as follows –

Air Resistance –

$$F_{ar} = C_d \times A \times rho \times v^2 / (2 \times g)$$

$$= C_d \times A \times 1.14 \times (V/3.6)^2 / (2 \times 9.81)$$

$$= 0.00483 \times C_d \times A \times V^2 \text{ .. (3.2)}$$

where, C_d – co-efficient of air drag
A - frontal area of the vehicle
Rho - density of air = 1.14 kg /m^3
v – vehicle speed in m/sec.
= V/ 3.6
– vehicle speed in kms/hr
g - gravitational acceleration
= 9.81 m / sec^2

Fuel consumption due to air resistance then can be calculated using eq^n (1.3) i.e.-

$$F_{Car} = C_f \times F_{ar}. \text{ ... eq}^n \text{ (3.3)}$$

For Model Car -
For the purpose of proper visualization, we have selected a model car with specifications as given in chapter 1.

Substituting the values of our model car in the eqⁿ (3.2) & eqⁿ (3.3), they reduce to –

$$Far = 0.00474 \times V^2 \ldots eq^n \ (3.2a)$$

$$FCar = 0.0038 \times V^2 \ldots eq^n \ (3.3a)$$

Using above equations force due to air resistance Far & fuel consumption due to air resistance FCar for various vehicle speeds are calculated & shown in the table 3.1.

Table 3.1
Air Resistance

For model Car -

V	20	40	60	80	100	120	140	160
Far	2	8	17	30	47	68	93	121
FCar	2	6	14	24	38	55	74	97

Far – Force due to air resistance

FCar – Fuel Consumption due to air resistance

In the last chapter on rolling resistance we had seen that

$$Frr = 16 \ kg.$$
$$\& \ FCrr = 13 \ cc/km$$

We had also seen that both Frr & FCrr are constant for given vehicle.

Adding above values of rolling resistance to values of air resistance given in Table 3.1, we get total values of F & FC due to (rolling resistance + air resistance) & the same are shown in the Table 3.2 –

Table 3.2
(Rolling Resistance + Air Resistance)

V	20	40	60	80	100	120	140	160
F(rr + ar)	18	24	33	46	63	84	109	137
FC(rr +ar)	15	19	27	37	51	68	87	110

Also graph of fuel consumption due to rolling resistance & air resistance is shown in Graph 3.1 below –

Graph 3.1
FCrr + FCar

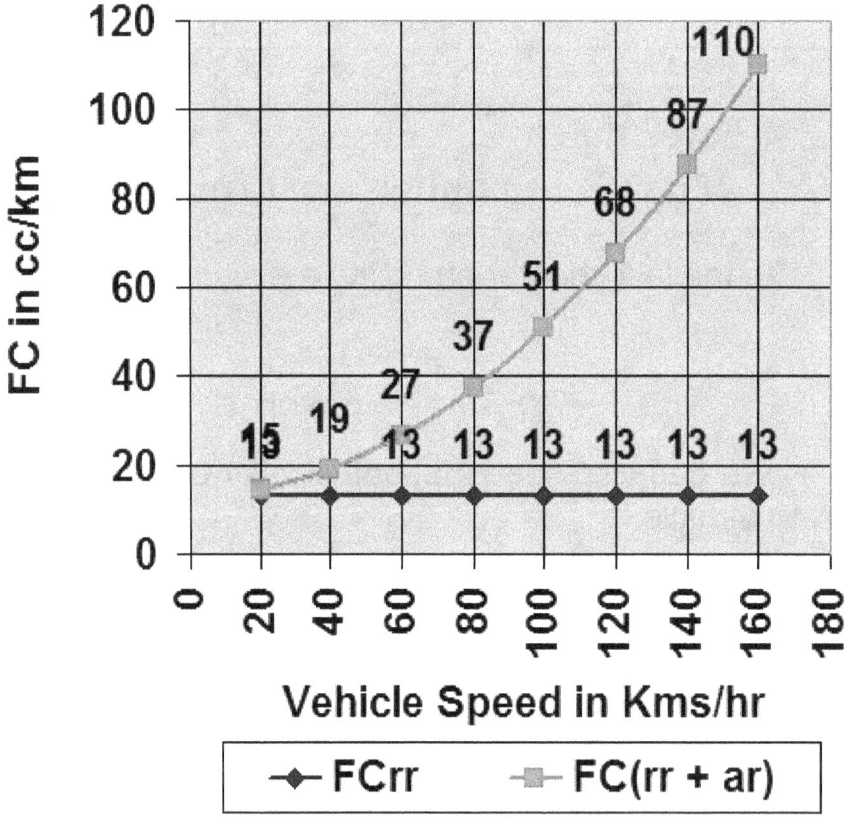

From the graph we can see how fuel consumption due to rolling resistance + air resistance goes on increasing rapidly with increase in vehicle speed.

Summary -

Fuel consumption due to air resistance varies in proportion to V^2, but it is independent of gear position.

---oOo---

4. Engine Friction

In the last chapter we have seen the energy requirements of the vehicle under different conditions. Now we will see how efficiently engine can supply us the energy required for driving the vehicle.

In case of any manufacturing plant, we get best possible plant efficiency if we operate the plant very near to 100 % load factor i.e. near to 100 % capacity of the plant. If for some reasons may be recession or any natural calamity or any other reason, if it becomes necessary to operate the plant at lower load factor, then plant efficiency will be low & production cost per unit of production will be high. This is because there are some fixed costs involved & this expenditure remains the same irrespective of the actual production. Therefore fixed cost per unit of production increases (e.g. if load factor is 50 %, then total fixed cost gets divided over 50 % of units & fixed cost per unit will be doubled) resulting in net increase in cost of production per unit.

In case of vehicles, engine is the power plant which is supplying the energy required for driving the vehicle & power output of the engine varies every moment. It varies from 100 % level to 0 % level The following examples will make the point clear –

i. When we are accelerating the vehicle at the maximum possible rate of acceleration, engine is giving maximum power output & load factor is nearly 100 %.

ii. But many a times due to traffic conditions on the road we are either accelerating at a very slow rate or just cruising. In this case load factor is low.

iii. When vehicle is stationary at the signal & engine is running, then load on the engine is zero.

iv. When we are running the vehicle with accelerator pedal released, in this case there is no fuel supply to the engine & engine is not producing any power. But engine is being driven using Kinetic Energy of the vehicle i.e engine is consuming power instead of giving any power output.

With variation in load factor of the engine there will be drastic variation in engine efficiency as load factor is varying from 0 % to 100 %.

The effects of this variation in engine efficiency can best be understood by applying the technique of fixed cost & variable cost as we do in case of any plant.

As long as engine is running at any given r.p.m. (revolutions per minute), it consumes some amount of fuel, even when it is not giving any net power output. This is the situation when the vehicle is stationary & engine is running. In this case power output is zero, but there is fuel consumption.

This fuel consumption is due to energy required to overcome internal friction of the engine. This fuel consumption is bound to be there all the time as long as the engine is running at that r.p.m. This part of fuel consumption represents fixed part of fuel consumption & is termed as 'No Load Fuel Consumption'.

When engine starts giving net power output as in the case of running vehicle, with increase in power output there will be increase in engine friction, but this part will be proportional to the increase in power output & is accounted in variable part of fuel consumption.

Thus total fuel consumption we have divided into two parts –

fixed & variable.

Fixed part is basically due to engine friction & is called as 'No Load Fuel Consumption' & variable part is proportional to load on engine.

Fuel Consumption due to Engine Friction – FCef -

No load fuel consumption of an engine at any given engine speed measured in r.p.m.s (**R**evolutions **P**er **M**inute) can be directly measured by keeping the vehicle stationary in neutral gear & running the engine at constant r.p.m. This no load fuel consumption is due to engine friction only. Accordingly, FC in liters/hr of our model car at different r.p.m.s is shown tabulated below –

Table 4.1
No Load Fuel Consumption

Engine R.P.M.	900	1500	2000	3000	4000
FC in liters/hr	0.29	0.58	0.87	1.63	2.60

Chart 4.1 shows variation in fuel consumption in liters/hr with engine r.p.m. based on the above table.

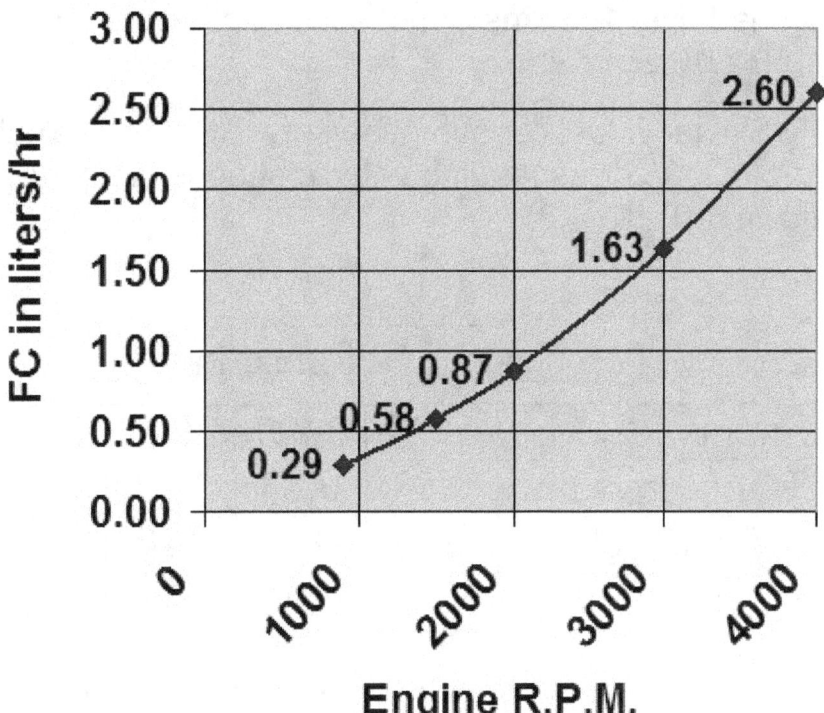

Graph 4.1
No Load FC in Liters/hr

For different engines, we will have different values of fuel consumption, but the basic pattern will remain the same.

It can be seen from the above table that with increase in r.p.m., increase in fuel consumption is more than proportion.

In any gear, as vehicle speed increases engine r.p.m. increases in direct proportion. But with increase in engine r.p.m. there is more than proportionate increase in fuel consumption as seen above. Therefore in any gear, with increase in vehicle speed there is more than proportionate increase in fuel consumption in cc/km.

Table 4.2 shows vehicle speeds in Top gear corresponding to different engine r.p.m.s. We have seen 'No load fuel consumption' of engine in liters/hr at different engine r.p.m.s. We will now see its effect on fuel consumption in cc/km.

At 4000 r.p.m. no load fuel consumption of engine is 2.6 liters/hr i.e. 2600 cc/hr. Vehicle speed corresponding to 4000 r.p.m. in top gear is 160 kms/hr. Therefore 2600 cc gets divided over 160 kms & fuel consumption/km – FCef = 2600/160 = 16.25 cc/km.

Similarly, fuel consumption/km – FCef are calculated for all r.p.m.s & corresponding vehicle speeds in top gear & are shown in Table 4.2.

Table 4.2
Top Gear - FCef in cc/km

Engine R.P.M.	1500	2000	3000	4000
FC in liters/hr	0.58	0.87	1.63	2.6
V in kms/hr	60	80	120	160
FCef in cc/km	10	11	14	16

From the Table 4.2 it can be seen that in Top Gear, fuel consumption due to engine friction – FCef goes on increasing from 10 cc/km to 16 cc/km.

Similarly, fuel consumption due to engine friction – FCef are calculated for all values of engine r.p.m.s in different gears & are shown in Table 4.3.

From table 4.3 it can be seen that for same engine r.p.m. of say 4000 r.p.m., in lower gears vehicle speed V gets reduced as per the ratio of gears of lower gear & top gear.

Therefore no load fuel consumption of 2600 cc/hr at 4000 r.p.m., gets divided over less no. of kms in lower gears. This results in higher value of FCef in lower gears. e.g. –

In 1st gear at 4000 r.p.m. vehicle speed is 32 kms/hr, therefore fuel consumption of 2600 cc/hr gets divided over 32 kms only (instead of 160 kms in top gear) & fuel consumption due to engine friction – FCef increases to 82 cc/km (instead of 16 cc/km in top gear)

Table 4.3
FCef - in cc/km in different gears

r.p.m.	1500	2000	3000	4000
FC_{NL}	0.58	0.87	1.63	2.60
V –kms/hr –				
Top	60	80	120	160
4th	46	62	92	123
3rd	32	43	64	86
2nd	21	28	41	55
1st	12	16	24	32
FCef in cc/km -				
Top	10	11	14	16
4th	12	14	18	21
3rd	18	20	25	30
2nd	28	32	39	47
1st	49	55	69	82

In any gear, as engine r.p.m. increases, FCef goes on increasing.

Therefore, we should avoid using higher engine r.p.m. range. Normally, for our model car, even though max. r.p.m. is 4000, it will be possible to shift to higher gear when r.p.m. is above 2600 r.p.m. & we can avoid higher engine r.p.m. range of 2600 to 4000 r.p.m.

Variation of fuel consumption due to engine friction in cc/km in different gears is shown in the graph 4.2 for better visualization.

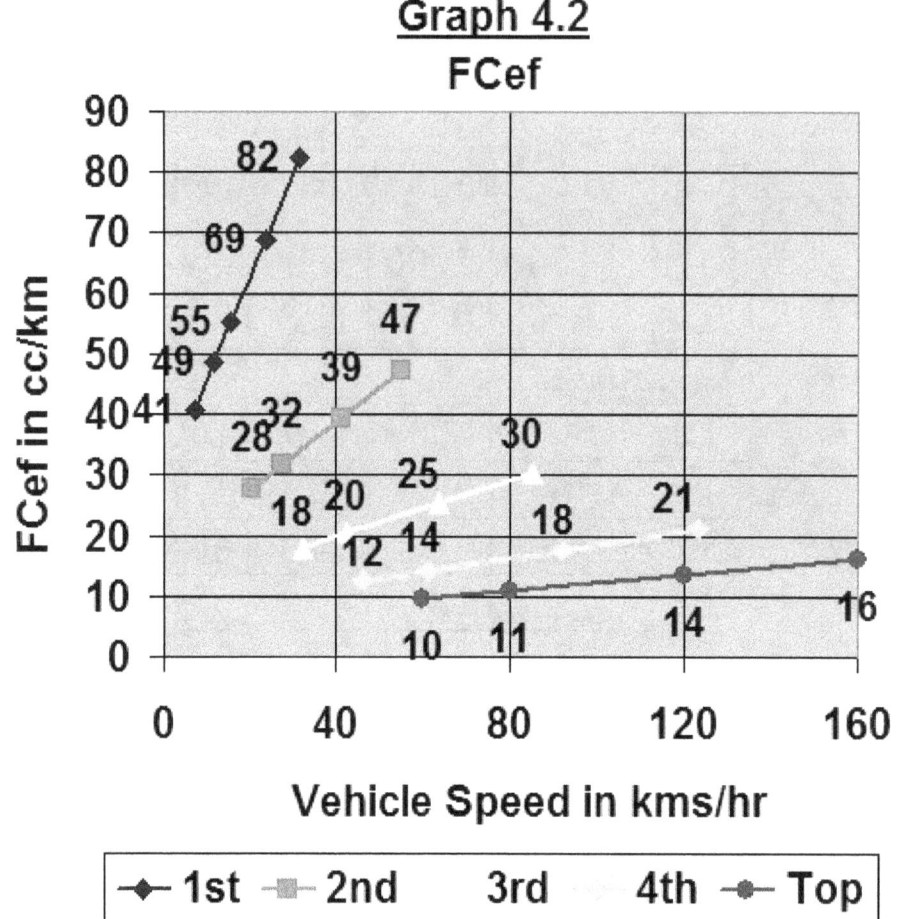

Graph 4.2

FCef

From the graph following observations can be made –

i. In any gear, as vehicle speed increases, FCef goes on increasing.

ii. In lower gears, FCef gets multiplied by gear ratio. Therefore FCef in lower gears is many times higher.

iii. In lower gears, with increase in vehicle speed, rise in FCef is much faster as compared to higher gear.

Therefore to reduce fuel consumption due to engine friction, we should always try to shift to higher gear early & avoid using higher engine r.p.m. range.

From equation 1.3 given in chapter 1, we know that –

$$FC = Cf \times F$$

Using above eqn, we can find value of force due to engine friction – Fef , from the value of FCef for all conditions given in Table 4.3. Values of forces are calculated accordingly & are shown in Table 4.4.

Value of force due to engine friction can be practically measured by very simple 'Retardation Test' as we will see in the chapter on 'Practical Testing'. Therefore values of forces due to engine friction are shown which will make it possible to compare theoretical values & practical values.

Table 4.4
Force due to Engine Friction – Fef

r.p.m.	1500	2000	3000	4000
FC_{NL}	0.58	0.87	1.63	2.60
V–kms/hr -				
top	60	80	120	160
4th	46	62	92	123
3rd	32	43	64	86
2nd	21	28	41	55
1st	12	16	24	32
Fef in kg -				
Top	12	14	17	20
4th	16	18	22	26
3rd	22	26	32	38
2nd	35	40	49	59
1st	61	69	86	103

Summary –

1. In any gear, FCef goes on increasing with increase in vehicle speed.

2. In lower gears, FCef gets multiplied by gear ratio of that gear & top gear.

3. In lower gears, with increase in vehicle speed, rise in FCef is much faster as compared to higher gear.

Therefore to reduce fuel consumption due to engine friction we should always try to shift to higher gear early & avoid using higher engine r.p.m.

---oOo---

5. Variations in Fuel Consumption

We have studied fuel consumption due to various factors i.e. –

i. FCrr – due to rolling resistance

ii. FCar – due to air resistance

iii. FCef – due to engine friction

Sum total of fuel consumptions due to these three factors in any gear & at any given speed gives us total fuel consumption at that constant speed on level road. i.e. –

$$FC = FCrr + FCar + FCef$$

Let us review what we have studied –

1. Rolling Resistance - We have seen that fuel consumption due to rolling resistance FCrr is given by the eq^n (2.2) & substituting the values for our model car -

$$FCrr = Cf \times (Crr \times W)$$
$$= 13 \text{ cc/km}$$

It can be seen that it is constant all the time & it is independent of vehicle speed & gear position.

2. Air Resistance – Fuel consumption due to air resistance of any given vehicle varies in proportion to square of vehicle speed & is given by the equation (3.3) i.e.

$$FCar = 0.00483 \times Cd \times A \times V^2$$

This equation is further reduced for our model car to equation (3.3a) –

$$FCar = 0.0038 \times V^2$$

It can be seen that FCar is proportional to V^2 only & is independent of gear in which vehicle is operated.

3. <u>Engine Friction</u> – For finding out the effect of engine friction, we had first measured fuel consumption in liters/hr at different engine r.p.m.s under no load condition & from that value we had find out fuel consumption in cc/km in different gears, for corresponding vehicle speeds.

We have seen that fuel consumption due to engine friction - FCef in any gear increases with vehicle speed more than proportionately. Also in lower gears FCef gets multiplied by gear ratio.

Thus FCef varies with both vehicle speed & gear position.

Sum of fuel consumptions for all the above 3 factors gives the total fuel consumption as given in eq^n (1.4) i.e.

$$FC = FCrr + FCar + FCef$$

Table 5.1
FC in cc/km & KMPL

r.p.m.	1500	2000	3000	4000
FC$_{NL}$	0.58	0.87	1.63	2.60
Top -				
V	60	80	120	160
FCef	10	11	14	16
FCrr	13	13	13	13
FCar	14	24	55	97
FC	36	48	81	126
KMPL	27.7	20.8	12.3	7.9

4th -				
V	46	62	92	123
FCef	12	14	18	21
FCrr	13	13	13	13
FCar	8	14	32	58
FC	33	41	63	91
KMPL	30.0	24.2	15.9	10.9

3rd -				
V	32	43	64	86
FCef	18	20	25	30
FCrr	13	13	13	13
FCar	4	7	16	28
FC	35	40	54	71
KMPL	28.8	24.9	18.6	14.1

2nd -				
V	21	28	41	55
FCef	28	32	39	47
FCrr	13	13	13	13
FCar	2	3	6	11
FC	42	47	59	72
KMPL	23.6	21.1	17	14.0

1st -				
V	12	16	24	32
FCef	49	55	69	82
FCrr	13	13	13	13
FCar	1	1	2	4
FC	62	69	84	99
KMPL	16.1	14.5	11.9	10.1

Accordingly, total fuel consumption & KMPL are calculated for different gears & different vehicle speeds & shown in Table 5.1.

Table 5.1 shows the following –

a. Values of vehicle speeds & FCef corresponding to different engine r.p.m.s in different gears (same as shown in Table 4.3).

b. Value of FCrr which is constant all the time - (13 cc/km).

c. FCar corresponding to vehicle speeds calculated using eq^n (3.3a) –

$$FCar = 0.0038 \times V^2$$

d. Total FC = FCrr + FCar + FCef

e. KMPL is calculated from FC using eqn (1.5) i.e. –
$$KMPL = 1000 / FC$$

Those interested in MPG figures can just multiply KMPL figures by 2.35. i.e. –

$$MPG = 2.35 \times KMPL$$

Using values of KMPL given in Table 5.1, graph is plotted & shown in Graph 5.1. It can be seen that –

i. KMPL of more than 25 can be achieved in speed range of 40 to 65 kms/hr. This is best KMPL range.

ii. At speeds above 120 kms/hr, KMPL is less than 13. (i.e. below 50 % of best KMPL)

iii. In lower gears, particularly if we operate in higher speed range of that gear, KMPL is below 15. (i.e. below 60 % of best KMPL)

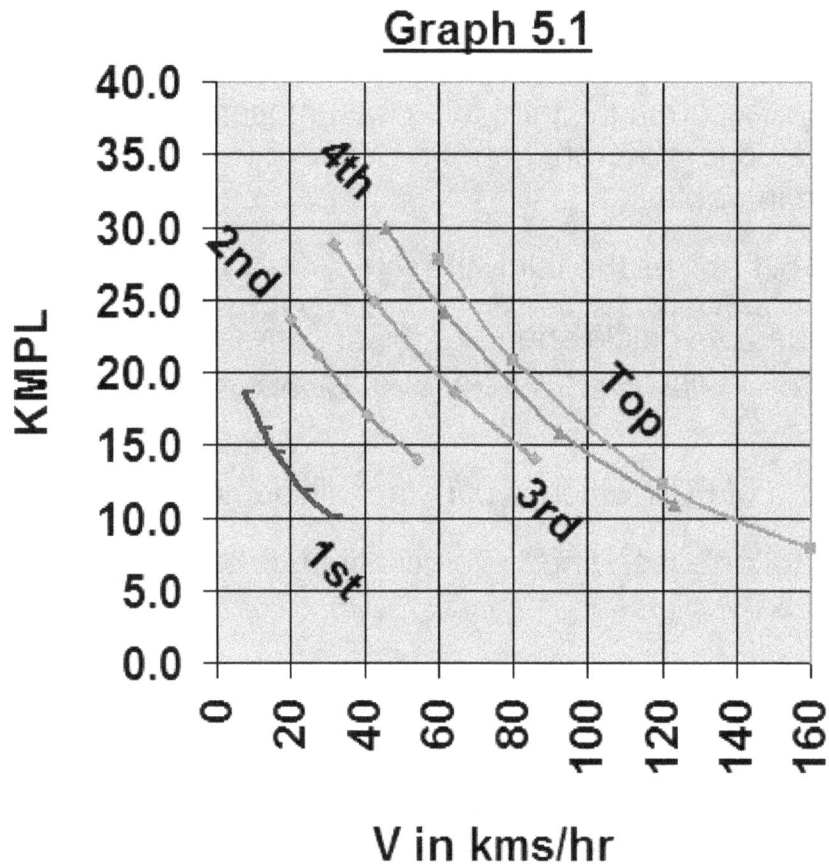

Practical verification of variations in KMPL –

1. If your vehicle is having MPG / KMPL meter, then above variations in different gears & different vehicle speeds can be easily checked. However care should be taken that vehicle is run at constant speed & on level road.

2. Otherwise, we can check the same by fitting a test pot & practically measuring fuel consumption.

3. Alternatively, one can compare MPG / KMPL under any two different conditions, by simply allowing vehicle to slow down through small speed range pre decided by you & measure time in seconds. Then ratio of time recorded in two tests will directly give you ratio of MPG / KMPL.

We will see all these methods in detail in next chapter on 'Practical Testing'. After all anything worked out theoretically needs to be confirmed practically.

5.1 Average Trip KMPL - From Table 5.1 & Graph 5.1 we can find out KMPL for running the vehicle in any gear & at any speed.

Even though we may operate the vehicle for most of the time in any gear & certain speed range, for achieving this speed vehicle has to be accelerated from 0 to the desired speed. For this vehicle is required to be operated in various low gears, starting from 1st, then 2nd & so on. In lower gears, vehicle will have low KMPL. Therefore, there will be drop in average KMPL depending upon % distance covered in lower gears.

Let us see what happens in different operating conditions –

5.1.A Highway Operation –

In highway running major distance will be covered in top gear at speed decided by us. If vehicle is operating in top gear at 80 kms/hr, for most of the time, then referring to Graph 5.1 average KMPL will be 21.

In Highway operation no. of stops or signals are comparatively low, therefore % distance travelled in lower gears will be very low, normally below 5 % or even below. Therefore drop in KMPL due to operation in lower gears is also low. But all depends upon the actual traffic conditions.

5.1.B City Operation –In city operation major portion of running will be around average speed of 30 to 50 kms/hr in 3rd gear. Average KMPL for this range is about 25.

But in City operations, no. of stops or signals per km, are very high. Naturally, after every stop, vehicle has to move through 1st gear, 2nd gear etc. Therefore % distance covered in lower gears is very high & therefore drop in KMPL is high. This considerably reduces KMPL in city operations.

After all our aim is to understand how KMPL is going to change under different conditions & how to get best possible KMPL in prevailing conditions.

We will see about this in subsequent sections.

5.2. Variations in Fuel Consumption – Even though expressing fuel consumption in terms of KMPL or MPG is normal practice now, expressing fuel consumption in cc/km has many advantages as given below –

i. % increase in fuel consumption in cc/km directly gives % increase in fuel cost/km.

ii. Break up of fuel consumption due to various factors i.e FCrr, FCar & FCef can be better visualized.

Therefore, graph of fuel consumption in cc/km is shown in Graph 5.2. (based on Table 5.1) In this graph apart from total fuel consumption in different gears, FCrr & (FCrr + FCar) are also shown in dotted lines.

We have already seen that FCrr & FCar are both independent of gear position. Only factor that is affected by gear position is FCef.

Therefore graph of (FCrr + FCar) is same for all gears. It is dependent only on vehicle speed. Naturally, vertical distance between graph of any gear & (FCrr + FCar) represents the value of FCef.

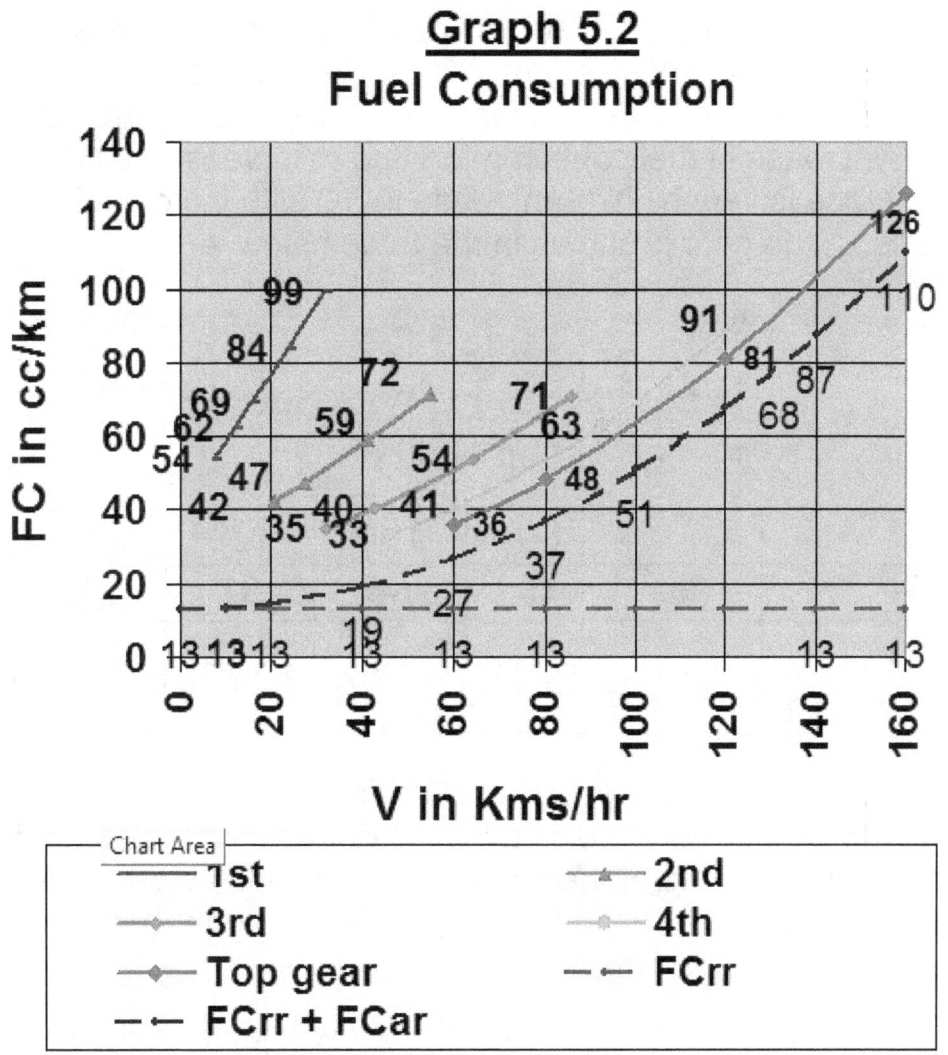

Graph 5.2
Fuel Consumption

From the graph 5.2 we can clearly see relative effect of each of the factor i.e. FCrr, FCar & FCef in total fuel consumption in any gear & at any vehicle speed.

We can clearly see that at higher speeds FCar is prominent, whereas in lower gears FCef is the most prominent factor.

Referring to Table 5.1 & Graph 5.2, following observations can be made –

5.2.A - In any gear, with increase in vehicle speed there is increase in fuel consumption.

In any gear, as vehicle speed increases there is increase in fuel consumption. This is due to increase in both FCar & FCef. Particularly in top gear, since speeds are very high, increase in fuel consumption due to increase in air resistance is very high. % increase in FC with increase in speed, in top gear is shown in the table below –

Table 5.2
% Increase in FC with Speed
(Top Gear)

V	60	80	100	120	160
FC	36	48	66	81	126
% increase	0 %	33 %	83 %	125 %	250 %

Even with increase in speed from 60 to 80 kms/hr fuel consumption increases by 33 % & if we increase speed from 60 to 100 kms/hr then increase is as high as 83 %. With further increase in speed % increase is still higher as shown in Table 5.2.

Therefore one has to take his own decision & have some practical compromise on vehicle speed for better fuel economy.

5.2.B - For the same vehicle speed, if we drive in lower gear fuel consumption is very high.

We have seen that fuel consumption due to rolling resistance + air resistance (i.e. FCrr + FCar) is independent of gear in which vehicle is driven & is dependent only on vehicle speed.

Therefore, for the same vehicle speed, fuel consumption due to rolling resistance + air resistance is same irrespective of gear in which vehicle is driven.

But FCef varies with gear & it is higher in lower gear. Let us see its effect on total fuel consumption.

Table 5.3
Effect of Operation in Lower Gears
(% Increase in Fuel Consumption)

Speed Kms/hr	Gear	Engine R.P.M.	FCef	FCrr + FCar	FC	% Increase
60	4th	1948	14	26	40	0 %
60	3rd	2805	24	26	51	26 %
40	3rd	1870	20	19	39	0 %
40	2nd	2909	39	19	58	49 %
25	2nd	1818	30	15	45	0 %
25	1st	3174	71	15	86	90 %

In Table 5.3 for each vehicle speed, for different gears, corresponding engine r.p.m.s, FCef, (FCrr + FCar) & FC are shown. It can be seen that for the same vehicle speed, in lower gears engine r.p.m. is higher which results in higher FCef. Therefore total fuel consumption FC is higher.

It can be seen that –

i. At speed of 60 kms/hr, if vehicle is driven in 3rd gear instead of 4th gear, increase in fuel consumption is <u>26 %</u>.

ii. At 40 kms/hr, if we use 2nd gear instead of 3rd then increase in fuel consumption is <u>49 %.</u>

iii. At 25 kms/hr, if we use 1st gear instead of 2nd gear then % increase in fuel consumption is <u>90 %</u>.

In City traffic we have no option but to drive the vehicle as per traffic on the road. Even if we want to run the vehicle in best fuel economy zone, it may not be possible. But at any vehicle speed, we have at least two options of selecting gear.

If we choose lower gear, we will be unnecessarily wasting more fuel to the tune of 26 % to 90 % as shown in the Table 5.2.

Therefore we should always drive in highest possible gear for better fuel economy.

5.2.C - Avoid using higher engine r.p.m. range – At higher engine r.p.m., fuel consumption due to engine friction is very high. Particularly in lower gears this effect is very much pronounced. Therefore we should avoid using higher engine r.p.m. range of the engine. Maximum engine r.p.m. of our model car is 4000 r.p.m., but we can shift to higher gear when engine r.p.m. is above 2600 r.p.m.

Graph of KMPL in different gears shown in Graph 5.1, is reproduced below with max. engine r.p.m restricted to 3000 r.p.m., as Graph 5.1a. Comparing this graph with original Graph 5.1, you can see that by restricting max. engine r.p.m. to 3000 r.p.m. we can avoid low KMPL area.

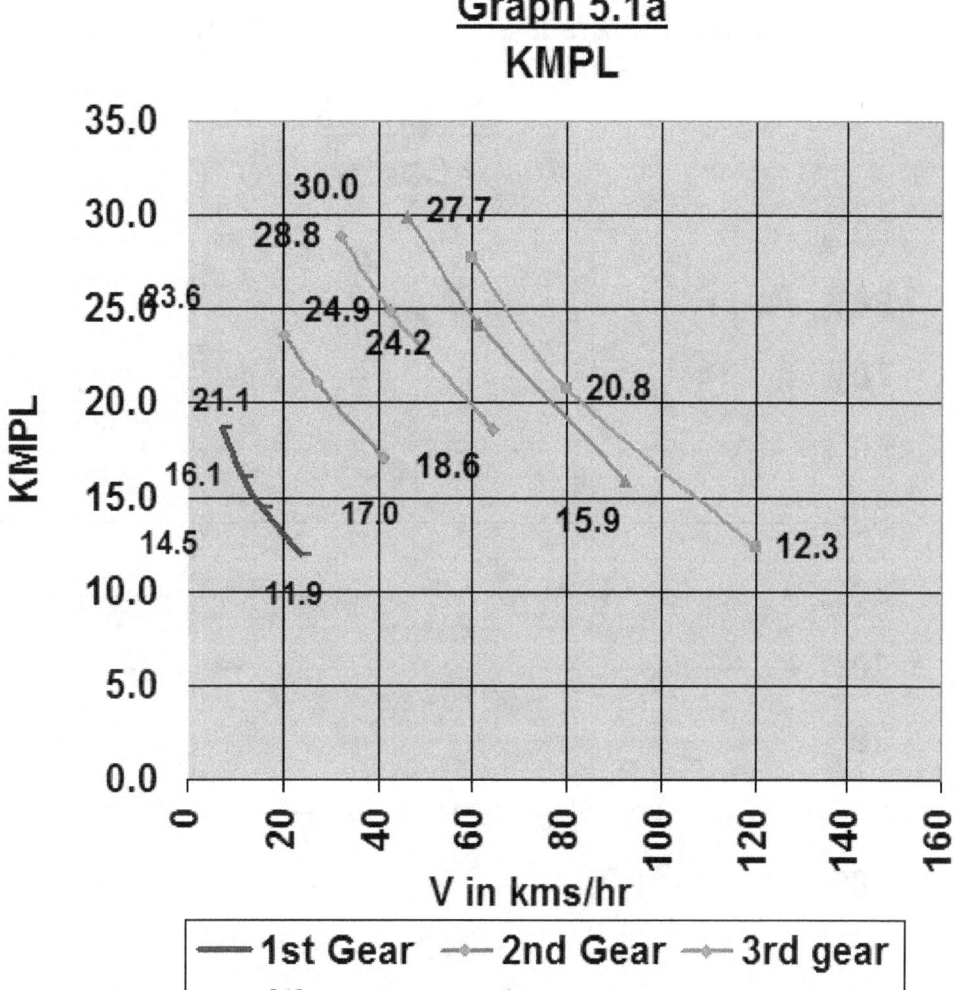

Graph 5.1a
KMPL

However, if we need more power as in case of climbing up a steep gradient or faster rate of acceleration then only we can use higher engine r.p.m.

Normally, those who drive heavy vehicles with front engines, can clearly hear engine noise & very well control engine r.p.m. once they understand fuel saving by this method.

5.3.A - % of Fuel Consumption due to Air Resistance in total Fuel Consumption –It is important for us to understand relative effect of different factors in total fuel consumption for better visualization. Therefore % of FCar in total FC in different gears is shown in the Graph 5.3.

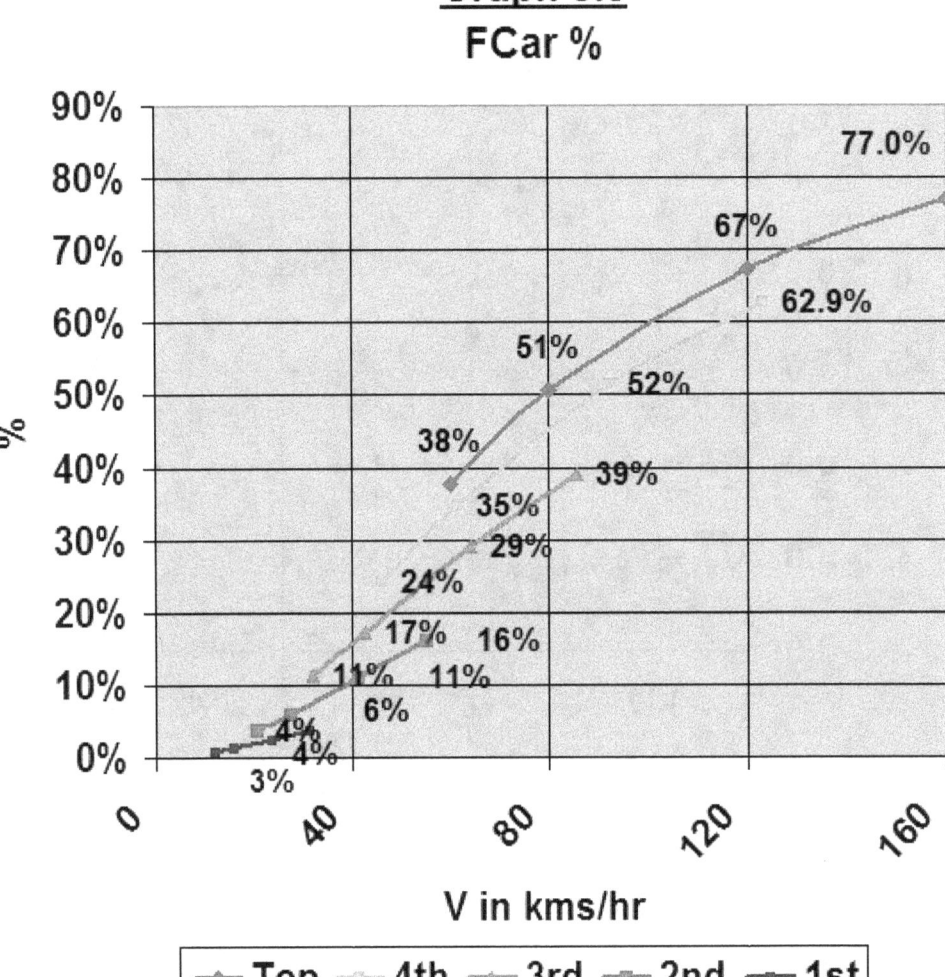

Graph 5.3
FCar %

It can be seen from the graph 5.3 that, above speed of 80 kms/hr, more than 50 % of fuel consumption is due to air resistance only & at still higher speeds it is more than 70 %.

Therefore obvious way of improving your fuel average is to keep control on vehicle speed while operating on highways.

If you are a vehicle designer, then you can think about reducing frontal area of the car & reducing co-efficient of air drag by improving aerodynamics of vehicle body to reduce fuel consumption due to air resistance.

5.3.B - % of Fuel Consumption due to Engine Friction - FCef, in total Fuel Consumption –

Similarly Graph of % of fuel consumption due to engine friction, in total FC, for different gears is shown in Graph 5.4.

In City operation, major portion of operation is at a speed less than 50 kms/hr. It can be seen from the graph that when vehicle speed is less than 50 kms/hr, % of fuel consumption due to engine friction varies from 50 % to 80 %. Naturally, if we have to improve KMPL in city operation, we must find ways to reduce FCef.

i. One of the obvious way is to drive in permissible highest gear for that speed. This is evident from the graph 5.2. We have already seen this in section 5.B.2 above.

ii. When engine is not supplying any power (e.g. when slowing down the vehicle), shift gear position to neutral –

Graph 5.4
FCef %

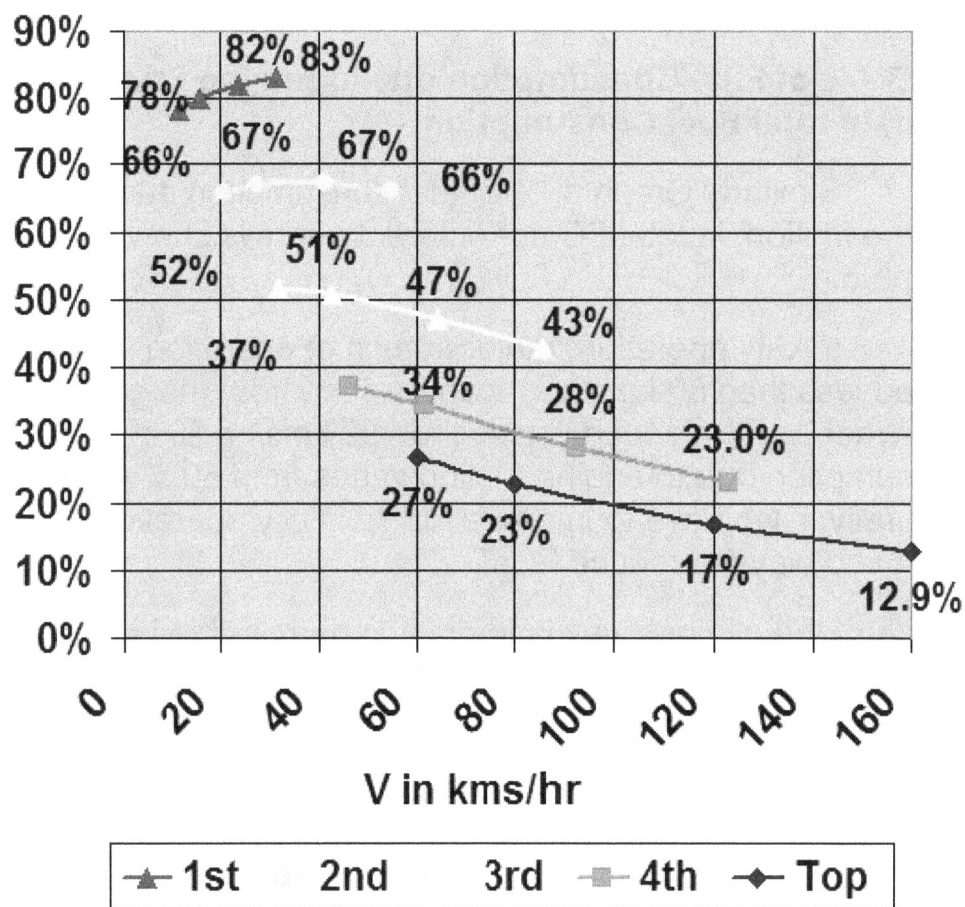

V in kms/hr

—▲— 1st   2nd   3rd   —■— 4th   —◆— Top

In neutral gear position, engine is disconnected from the rest of transmission. Therefore there is no load of engine friction on the vehicle. Naturally, there is reduction in energy required for driving the vehicle to the tune of 50 % to 80 % as seen from the graph 5.4.

When vehicle running at speed, is allowed to slow down by releasing accelerator pedal, there is no fuel consumption, but vehicle is able to cover considerable distance using kinetic energy of the vehicle. In this condition if vehicle is in neutral then there will be saving of 50 % to 80 %. Let us say 60 %. It means that we will be consuming only 40 % of energy & vehicle will be able to cover 2.5 times the distance which vehicle would have covered in that gear position.

Even though shifting to neutral results in lot of saving in energy, this method needs be used with due care. Problems are as follows –

i. If it becomes necessary to re-engage gear before stopping the vehicle then synchronizing with gear box out put shaft speed is difficult particularly when vehicle speed is high.

Therefore this technique of shifting to neutral while slowing down the vehicle should be used in such a way that it will not be necessary to re-engage the gear before stopping the vehicle when speed is high. In case if it becomes necessary to reengage gear when speed is high, then it is better to stop the vehicle than taking the risk of damaging the gearbox.

ii. If vehicle is moving down a steep gradient & for longer distance, then there is a risk of vehicle over speeding & going out of control. Also, if vehicle speed is controlled by continuous braking, it can result in over heating of brake liners & may cause brake fading. This reduces brake efficiency drastically & can lead to dangerous situation. After all we must keep in mind that safety of your vehicle & you are more important than fuel economy. Therefore do not shift to neutral while moving down the steep gradient.

Therefore be cautious & use this technique of slowing down in neutral with due care.

We have seen in this chapter fuel consumption at constant speed running on level road. But it is not final fuel average.

We have seen that there is no need to account for extra fuel required for acceleration or moving up the gradient. But we need to add fuel consumption corresponding to braking losses & idling losses.

We will see about all these remaining points in subsequent chapters.

Summary –

1. Graph 5.2 of fuel consumption in different gears along with graphs of FCrr & (FCrr + FCar) gives complete picture of how fuel consumption varies. This graph also gives the break up of fuel consumption i.e. FCrr, FCar & FCef at a glance.

2. Best economy zone is in the range of 40 kms/hr to 65 kms/hr.

3. In any gear, with increase in vehicle speed there is increase in fuel consumption & % increase can be 100 % or 200 % or even higher at higher speeds.

4. In lower gears, there is increase in fuel consumption due to increase in engine friction & % increase can be 50 % or even more than 100 %.

In this book we will be discussing only about variations in fuel consumption of our model car for avoiding confusion if we start discussing various types of vehicles. Once we understand variations in any one type of vehicle, we can easily understand variations in other types of vehicles. However in appendix we have given sample analysis for model bus. In the analysis Table similar to Table 5.1 & Graph of KMPL similar to Graph 5.1 is given. Method used is same as used for model car. Only difference is in the actual values for different types of vehicles.

This method & formula used can be used for any type of vehicles & any type of fuel.

---ooo---

6. Practical Testing

Up till now we have studied how fuel consumption varies under different conditions of vehicle speeds & in different gears. For practical verification of all the above points, one can measure fuel consumption while driving the vehicle under specified condition of vehicle speed & gear.

Normally, most of the people have a tendency to check fuel consumption for a complete trip. There is strong belief that if longer test is conducted result will be more accurate.

However we have seen that variations in fuel consumption under different conditions are very large, extending to even 200 %. Therefore even with, slight change in operating condition in part of the trip, can change fuel average for the trip (MPG/KMPL) to considerable extent & we do not get consistent readings. In order to get consistent readings it is better to test for a short duration so that vehicle can be run exactly at specified speed without any disturbance & particularly with absolutely no braking (not even keeping your foot on brake pedal).

Also testing should be done on level road. If you have doubt about the slope, you can perform two tests in opposite directions & take average to eliminate any effect of gradient.

Following methods can be used for practically checking fuel consumption under different conditions –

6.1 Using MPG / KMPL meter provided on vehicle –

If your vehicle is fitted with MPG / KMPL meter, then you can very well check it on the meter by running the vehicle at constant speed in any gear. It should be remembered that any variation in speed or even slight gradient of the road can effect your reading considerably.

Therefore testing should be for very short duration, but exactly at specified conditions.

Following tests can be performed –

1. MPG / KMPL meter can be used for checking fuel average at constant speed & on level road.

2. This meter can also be used for measuring drop in MPG / KMPL for the same speed as above for various levels of accelerations or on different gradients. But do not forget that drop in MPG / KMPL do not affect fuel average for the trip.

3. You can also check MPG / KMPL when you simply allow the vehicle to slow down just by releasing the accelerator pedal. You will find that meter is showing MPG / KMPL very high or beyond range. Real fact is that practically there is no fuel consumption.

If MPG / KMPL meter is not provided on your vehicle then still you have following alternatives -

6.2- Fuel Consumption Measurement –

Aim – To measure fuel consumption in cc/km for any specified condition.

Apparatus –

i. Test pot of 100 cc with markings of 1cc.

ii. 3 way valves – 2 no.s
   (either manual or solenoid operated)

iii. fuel pipes etc.

iv. stop watch

Set up for experiment – (Pl. refer Fig. 6.1)

i. In fuel supply line, fit 3 way valve & test pot in such a way that fuel supply to engine can be made either from regular tank of the vehicle or from test pot.

ii. In fuel return line from engine to tank, fit another 3 way valve in such a way that fuel from engine can get returned to the tank of the vehicle or can go to test pot.

Lay out of fitment of 3 way valve & test pot is shown in the Fig. 6.1 for ready reference.

Normally, fuel consumption measurement is done by fitting a test pot. But fuel from test pot is used for both test running as well as normal running of the vehicle. Therefore test pot needs to be of bigger size & naturally, min. quantity of fuel measured is in liters.

Whereas with the arrangement as shown in Fig.6.1, since fuel from test pot is used only during test, we can have much smaller test pot & fuel measurement in cc is possible. This reduces both test time & fuel for testing.

Reducing test time is of utmost importance to us because when we are testing the vehicle on public road, it is extremely difficult to run the vehicle exactly at constant speed without any disturbance.

Fig. 6.1
Arrangement for
Fuel Consumption Measurement

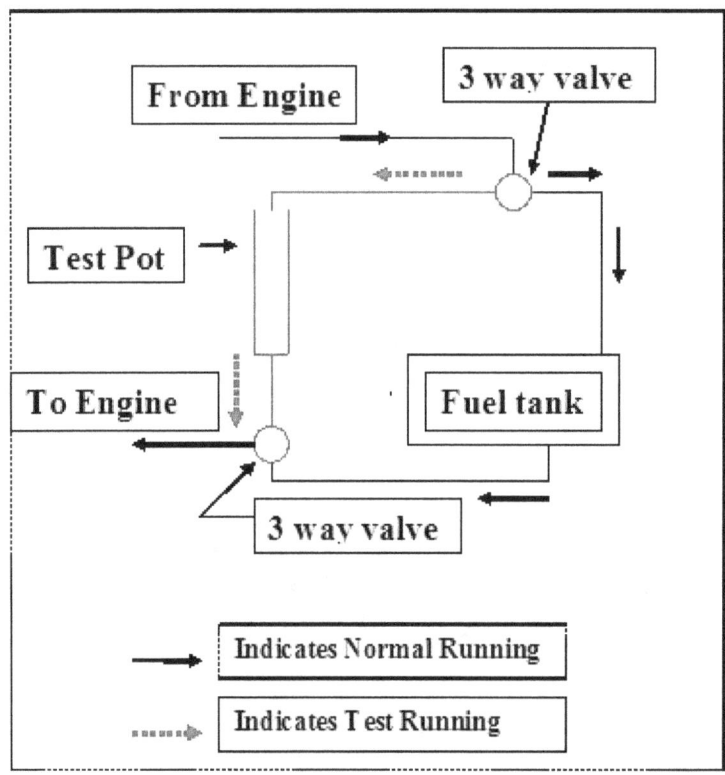

Testing –

1. First run the vehicle at desired speed using fuel from fuel tank & fuel return also going to tank.

2. When desired speed at which vehicle is to be tested is achieved, then we can operate both 3 way valves so as to take fuel supply from test pot & diverting fuel return coming from engine to test pot.

3. Run the vehicle exactly at required speed on level road. Note the fuel reading on test pot & start stop watch. Allow it to run till you consume about say 25 cc of fuel & note time required to consume 25 cc of fuel in seconds up to 1 or 2 decimal places.

Fuel consumption in cc/km can be calculated using the formulae –

$$FC = 'X' \times 3600 / (V \times t) \ldots (6.1)$$

Where , X – fuel consumed in cc.

(even though we have recommended to take 25 cc for cars, for heavy vehicles we will have take more quantity & also for checking conditions with high fuel consumption we can take reading for more fuel quantity for improving accuracy if traffic condition on road permits to take longer distance test without any disturbance)

Since with the above arrangement we are practically checking instantaneous fuel consumption, this method can be used even for checking fuel consumption during acceleration or while moving up a gradient.

Normally we will be interested in knowing fuel consumption in the following conditions –

a. At vehicle speed of around 60 kms/hr speed in top gear when we are getting best fuel average i.e lowest fuel consumption.

b. At V= 100 or 120 kms/hr in top gear to find out % increase in fuel consumption compared lowest fuel consumption noted above.

c. In lower gear, at some specified speed to find out % increase in fuel consumption compared to lowest fuel consumption in top gear.

Arrangement as above provided on vehicle is very useful even for training of drivers. We can practically demonstrate to the drivers various points that affect fuel consumption.

In addition to the above points given in a,b & c above, we can also demonstrate to drivers the following points –

i. For any vehicle speed instead of operating in higher gear, if we continue to operate in lower gear, what is increase in fuel consumption. (This increase in fuel consumption can be in the range of 25 % to 90 % as shown in Table 5.3)

ii. We can practically demonstrate to drivers that when we are running the vehicle with accelerator pedal released there is practically no consumption of fuel. It means that distance covered with accelerator pedal in released position is without consuming any fuel.

iii. We can also demonstrate to drivers, fuel consumption of engine under no load condition (i.e. with vehicle stationary) at various engine r.p.m.s. At higher engine r.p.m.s fuel consumption is abnormally high. In lower gears, engine r.p.m. is high as compared to higher gear for the same vehicle speed that is why fuel consumption is very high in lower gears.

Giving practical demonstration to drivers about the above points makes dramatic change in the driving habits of drivers.

This technique was used by me while working in a large public transport company for training 10000 drivers & improvement in average KMPL for the fleet of more than 3500 buses was almost 20 %. Educating masses & making them to implement it, is not an easy task. But in this case, drivers were able to see with their own eyes, how much extra fuel they are consuming due to wrong practices & this resulted in very good response from the drivers.

In fact, it was observed that there were many drivers who improved their KMPL by more than 30 %.

Even though above method of actual fuel measurement is very effective method for understanding how KMPL is changing in different conditions, it can be done by fleet operators or vehicle manufacturers only. But individual vehicle owner cannot afford to do all this.

For them, carrying out 'Retardation Tests' under different conditions of vehicle speeds & gears is a very effective method for understanding variation in MPG / KMPL under different conditions. Same is explained below –

6.3 - Retardation Test – We have seen that force required to drive the vehicle is sum total of opposing forces like rolling resistance, air resistance & engine friction. We have seen how these forces vary under different conditions.

When engine is driving the vehicle, engine is supplying this force. But when engine is not supplying power, still this force continues to act on the vehicle & there is retardation of the vehicle under the action of this force.

When the vehicle is running at high speed & we release the accelerator pedal then there is no supply of fuel to the engine & engine is not giving any force to the vehicle. Therefore only force acting on the vehicle is the total force due to rolling resistance, air resistance & engine friction. This force gives retardation (-ve acceleration) to the vehicle & vehicle speed goes on reducing. Rate of retardation is given by –

Force = mass x retardation .. (6.2)

Therefore by simply measuring the rate of retardation we can find the value of force at any given speed & specified gear.

For checking the rate of retardation in any gear at any vehicle speed V, we need to observe time 't' required for certain drop in vehicle speed by say 20 kms/hr [i.e. from (V + 10) to (V - 10)]. Then –

Drop in speed

= 20 kms/hr
= 20 x 1000 / 3600 m/sec.
(Since 1 km = 1000 meters
& 1 hr = 3600 sec.s)
= 20 / 3.6 m/sec.
= 5.56 m/sec.

Then we can calculate the force acting on the vehicle at V kms/hr in any gear by using the formulae –

Force
= mass x retardation
= (W/g) x [(drop in speed)/t] .. (6.2)
= 0.566 x W / t   .... (6.2a)
(above formulae is for drop in speed of 20 kms/hr)
where,
W– total weight of the vehicle
= 1600 kg for model car
g - gravitational acceleration
= 9.81 m/sec^2.
t - time in seconds

Using the above formulae (6.2) we can find the value of force at any given speed & any specified condition.

Above formulae for force can be used for any vehicle right from scooter, motor cycle, 3 wheelers, cars or buses or trucks.

6.3.A –Calculating fuel consumption from Force F measured during Retardation Test –

Using eqn (1.3) we can find fuel consumption by just multiplying force F measured during retardation test by conversion factor Cf. i.e. –

FC = Cf x F

Therefore substituting value of force from formulae (6.2a) –

FC

$= Cf \times F$

$= Cf \times 0.566 \times W/t \dots (6.3a)$

Also using formulae (6.3a) we can calculate fuel consumption, but for this we need to know the value of conversion factor Cf. We can use assumed value of Cf or practically check the same.

For practically measuring Cf, after conducting retardation test for any one set of condition & calculating value of force, we can physically measure fuel consumption for the same conditions & then calculate value of Cf.

Even if there is some error in assuming estimated value of Cf , still it will give clear picture of how fuel consumption varies under different conditions. Since FC is proportional F, % variation in fuel consumption will remain the same irrespective of value of Cf.

6.3.B – Comparing MPG / KMPL under different conditions of vehicle speeds & gears –

In 6.3.A we have seen the method of measuring force & fuel consumption by carrying out retardation test, but actually we are more interested in knowing under what condition we get best KMPL & compared to that what is the % drop in MPG / KMPL in any other condition.

From formulae (6.3a) it can be seen that fuel consumption FC is inversely proportional to time 't' measured during retardation test.

Therefore, for comparing fuel consumption for any two different sets of conditions we can just carry out retardation test & measure time - 't' for those two sets of conditions.

If time recorded are t_1 & t_2, for set1 & set 2 respectively, then –

$$FC_2 / FC_1 = t_1 / t_2 \ldots (6.4)$$

Further, we know that KMPL = 1000/ FC,

i.e. KMPL is inversely proportional to FC.

Therefore –

$$KMPL_1/KMPL_2$$
$$= FC_2/FC_1$$
$$= t_1/t_2 \ldots (6.5)$$
[using eqn (6.4)]

From above equation (6.5), we can see that **KMPL is directly proportional to time 't' recorded during retardation test**. Since MPG =2.35 x KMPL, same formulae also can be used for comparing MPG. i.e. –

$$MPG_1/ MPG_2 = t_1 / t_2$$

This method can be used for any type of vehicle & need no other data of the vehicle for comparing fuel consumptions or MPG / KMPL.

These tests need absolutely no instruments or special arrangement except stop watch to measure time which is now a days is available in our mobiles also. Since each test requires only few seconds, for all the tests under different conditions takes only few minutes.

<u>This method is very effective tool for individual vehicle owners to compare MPG / KMPL under different conditions.</u>

6.3.C – Finding out Frr, Far & Fef , Crr, Car –

Engineers involved in R & D, will like to know the values of Frr, Far & Fef separately & also to find out values of Crr, Cd & Cf to assess the effect of any changes they make in the design.

By carrying out retardation tests under different conditions we can easily find out the values of Frr, Far & Fef & from these values we can find out the values of co-efficient of rolling resistance – Crr & co-efficient of air resistance – Cd easily. Otherwise finding out Crr & Cd is not an easy task.

For measuring the values of Frr, Far & Fef we have to conduct following tests –

Test – Carry out retardation tests for four different conditions as given in Table 6.1 below & measure time 't' for each of the condition for vehicle speed to drop down by say 20 kms/hr.

Calculate value of force for each condition using formulae (6.2).

Tabulate the results as given in Table 6.1.

Table 6.1
Natural Retardation Tests

Test no.	Speed	Gear	Time	Force
1	V_1	Top	t_1	F_1
2	V_1	Neutral	t_2	F2
3	V_2	Neutral	t_3	F3
4	V_2	Top	t_4	F4

From the value of forces F_1, F_2, F_3 & F_4 we can easily work out the values of Frr, Far, Fef & also Crr & Car as follows –

6.3.C.a – Force due to Engine Friction – Fef –

We know that value of force – F calculated from time measured during retardation test is given by –

F = Frr + Far + Fef

We also know that –

Frr – is constant for all vehicle speeds & gears

Far – proportional to V^2

& Fef – depends upon V & gear ratio.

Therefore it is better to use proper suffix to specify condition to avoid confusion. e.g. For condition 1 in above table –

$$F_1 = Frr + Far_{(V1)} + Fef_{(V1, top)} \dots (6.6)$$

(Pl. note that for Frr no suffix is necessary, for Far suffix to indicate speed is added & for Fef suffix indicates both speed & gear)

In test condition 2, we have used neutral gear. In neutral, there is no connection between engine & wheels. Therefore there is no load of engine friction on the vehicle & Fef will be zero.

Therefore –

$$F_2 = Frr + Far_{V1} \dots (6.7)$$

(In this test, since speed is same as test 1 i.e. V_1, Far will be just the same i.e. $Far_{(V1)}$)

From equations (6.6) & (6.7) –

$$F_1 - F_2 = Fef_{(V1, top)} \dots (6.8)$$

We can see that for finding out value of Fef at any given speed & in any gear, we have to carry out two retardation tests first at specified speed & gear & then at that speed but keeping gear position in neutral. Difference in forces calculated from two tests directly gives us the value of Fef at that speed & gear.

6.3.C.b. <u>Force due to Air Resistance – Far –</u>

From test 2 & 3 we get –

$F_3 - F_2$
$= [Frr + Far_{V2}] - [Frr + Far_{V1}]$

$= Far_{V2} - Far_{V1}$ .. (6.9)

But since, Far is proportional to V^2,

$Far_{V2} = (V_2/V_1)^2 \times Far_{V1}$

Substituting in equation (6.9) we get –

$F_3 - F_2$

$= (V_2/V_1)^2 \times Far_{V1} - Far_{V1}$

$= [(V_2/V_1)^2 - 1] \times Far_{V1}$ .. (6.10)

Using eqn (6.10) we can calculate value of Far_{V1}.

Once we know the value of Far at V_1, we can calculate value of co-efficient of air drag - Cd using eq^n (3.2) given in chapter 3 on 'Air Resistance'. i.e. –

$Far = 0.00483 \times Cd \times A \times V^2$

6.3.C.c Rolling Resistance – Frr –

Further substituting the value of Far_{V1} in equation (6.7), we can calculate value of Frr & from the value of Frr we can also calculate value of co-efficient of rolling resistance Crr using eqn (2.1) given in chapter 2 on 'Rolling Resistance' i.e.

$Frr = Crr \times W$

6.3.C.d - To Find Out Conversion Factor Cf –

We have already measured force at two different speeds in top gear as given in Table 6.1. Also we have seen the method of measuring fuel consumption in 6.2.

Using this method we can measure fuel consumption in cc/km for test condition 1 or 4 given in Table 6.1 & then calculate value of Cf using eq^n (1.3) i.e. – FC = Cf x F

Or alternatively, since we have already calculated Force due to engine friction Fef at speed V_1, we can measure no load fuel consumption of engine at r.p.m. corresponding to speed V_1. Then you can calculate value of Cf using equation (1.3)   i.e. FC = Cf x F. i.e. –

Cf = FCef / F
     = [(no load FC x 1000)/ V_1] / F

From all the above, it can be seen that just by carrying out few retardation tests, we can find values of all forces acting on the vehicle i.e. Frr, Far & Fef. Also we can calculate the values of Crr & Cd using these values of forces. Since these retardation tests can be carried out without any instruments & can be carried out just in few minutes, these test are very useful for understanding variations in fuel consumptions for individual vehicle owners.

---oOo---

7. Acceleration

We have so far seen the total force required to overcome rolling resistance, air resistance & engine friction for running the vehicle at any constant speed, in any gear & corresponding fuel consumption.

We have seen in the first chapter on 'Introduction' that even though extra fuel is consumed during acceleration there is no need to account for this extra fuel for acceleration for trip MPG / KMPL. Let us see the details of the same.

When we are accelerating the vehicle, with increase in vehicle speed, there is an increase in 'Kinetic Energy' of the vehicle. For providing this increase in energy, extra fuel is required.

But when we are reducing speed this energy is released & we are able to cover extra distance using this energy.

7.1 Extra Fuel Required for Acceleration -When we are accelerating the vehicle, with increase in vehicle speed, there is increase in kinetic energy of the vehicle. Therefore extra fuel is consumed for providing this energy.

In this case there is no energy consumption. But energy produced is stored in the vehicle in the form of kinetic energy.

Kinetic energy possessed by vehicle at any speed V is given by the equation –

$$E$$
$$= \frac{1}{2} \times m \times v^2. \ldots$$
$$= \frac{1}{2} \times (W/g) \times (V/3.6)^2 \ldots eq^n (7.1)$$
Where,
m – mass of vehicle
$$= (W/g)$$
W = weight of vehicle
v = vehicle speed in m/sec.

V = vehicle speed in kms/hr

Using eq^n (7.1) values of K.E. are calculated for different vehicle speeds for our model car & are shown in Table 7.1. (value of E calculated using eq^n (7.1) are in kg-m, but they are then converted into kg-km by dividing with 1000 for the sake of convenience)

Also quantity of fuel required in cc for producing these energies are calculated using formulae (1.2) i.e. FC = Cf x E & are shown in Table (7.1)

Table 7.1

Extra Fuel for Kinetic Energy

V in kms/hr	K.E. In kg-km	Quantity of fuel In cc
40	10	8
50	16	13
60	23	18
80	40	32
100	63	50
120	91	72
140	123	99
160	161	129

From the above table we can see that when our model car is running at speed of 60 kms/hr, kinetic energy possessed by vehicle is 23 kg-km & quantity of fuel required to provide this energy is 18 cc.

It means that for accelerating the vehicle from 0 to 60 kms/hr speed, vehicle will consume 18 cc of extra fuel i.e.

in addition to normal fuel consumption for overcoming rolling resistance, air resistance & engine friction as studied earlier.

If we further accelerate the vehicle from 60 kms/hr to 80 kms/hr, then increase in K.E. will be 40 − 23 = 17 kg-km & extra fuel required will be 32 − 18 = 14 cc.

From the above table we can find out the increase in K.E. (kinetic energy) & quantity of fuel in cc for any speed range.

7.2 When Speed is Reduced, Kinetic Energy is Released & compensates extra fuel during Acceleration –

We have seen in 7.1 above that when vehicle speed is increased through certain speed range, extra quantity of fuel is consumed corresponding to increase in kinetic energy of the vehicle. This was shown in Table 7.1

But when vehicle speed is reduced through same speed range, same amount of kinetic energy is released & we can cover extra distance using this kinetic energy.

When we release accelerator pedal & allow the vehicle to slow down, fuel supply to the engine is cut off. In this case, vehicle is running using only kinetic energy of the vehicle. In short we are able to cover extra distance using kinetic energy only & without consuming any fuel.

When vehicle slows down, energy required for running the vehicle is the energy required to overcome rolling resistance, air resistance & engine friction, as seen earlier. This energy is supplied by release of kinetic energy of the vehicle.

In normal running, we are consuming fuel & covering some distance immediately.

In case of acceleration, extra fuel is consumed during acceleration, but its output in terms of extra distance we are getting when vehicle speed is reduced. It is like money deposited in bank for future use.

Moreover, energy produced by consuming extra fuel during acceleration, is ultimately used for running the vehicle as in normal course i.e. for overcoming rolling resistance, air resistance & engine friction. Therefore out put of extra fuel i.e. distance covered will be the same as in normal course.

<u>Therefore there is no need to account for extra fuel consumed during Acceleration for calculating Trip KMPL</u>.

However, we need to remember following points –

i. when we are accelerating the vehicle fuel consumption in cc/km will be more & MPG/KMPL will be low.

ii. But when speed is reduced vehicle will be able to cover more distance using kinetic energy released, without consuming any fuel & MPG/KMPL will be very high or infinity. (since fuel consumption is practically zero)

iii. Overall MPG/KMPL for the trip will remain unchanged as energy produced by extra fuel consumed during acceleration is utilized when vehicle speed is reduced as in normal running.

7.3 Faster Acceleration Reduces KMPL <u>during acceleration</u>, but Trip KMPL remains unaffected –

We have already seen that extra fuel is consumed for acceleration of the vehicle depending upon the speed range, but there is no need to account for this extra fuel as seen in 7.2 above.

We have seen that for accelerating the vehicle extra fuel is consumed depending upon the speed range as shown in the Table 7.1. e.g. for accelerating the vehicle from say 60 kms/hr to 80 kms/hr we will need 32 – 18 =14 cc of extra fuel.

If we accelerate the vehicle at a faster rate, then this 14 cc of fuel will be consumed in shorter distance & extra fuel consumption due to acceleration – FCa expressed in cc/km will be higher.

But if we accelerate at a slower rate, then this 14 cc of fuel, will be covered in more distance & therefore FCa expressed in cc/km will be less.

From the above we can see that even though fuel consumed in both the above cases is just the same, with faster acceleration we will have higher value of FCa. Therefore % increase in FC with faster acceleration will be much higher. In fact, it can be more than 150 % also.

But the fact remains that there is no energy consumption & this energy remain stored in the vehicle & vehicle covers extra distance using this energy when we slow down the vehicle.

Therefore as seen earlier there is no need to account for extra fuel for acceleration for calculating Trip KMPL as speed at the start & end of the trip is zero & there is no change in kinetic energy of the vehicle at the end of the trip.

[Only care we have to take while accelerating in lower gears is that if rate of acceleration is much faster than holding capacity of the tyre, then it can result in slippage of driven tyres that can cause energy loss for short duration. Except this, it is beneficial to accelerate at a faster rate.]

7.4 Faster Acceleration in lower gears Improves Trip KMPL –

Some people are advocating that 'Do not accelerate at faster rate as it increases fuel consumption & reduces KMPL'. **But it is only half truth.**

From all the above analysis it can be seen that faster acceleration reduces KMPL during the period of acceleration only, but it is compensated by running the vehicle for extra distance without any fuel when we slow down the vehicle & overall there is absolutely no effect on trip KMPL.

In contrast to this actually it is better to accelerate the vehicle at faster rate when we are in lower gears as explained below –

We have seen that when we are operating in lower gear, there is considerable increase in fuel consumption, due to increase in fuel consumption due to engine friction - FCef.

Therefore if we are accelerating at a slower rate of acceleration, then we will be spending more time in lower gears & we will be unnecessarily spending more fuel.

As against this, if we accelerate at a faster rate, we will be shifting to higher gear early & saving considerable amount of fuel. (We have seen in the chapter on 'variations in fuel consumption' -Table 5.2 that operation in just one gear lower, increases fuel consumption by 26 % to 90 %.)

Therefore when we are in lower gears, it is economical to accelerate at a faster rate so that we can shift to higher gear early for better fuel economy.

7.5 Extra fuel consumed during acceleration is like deposit in the bank & can earn interest too –

Vehicle is first accelerated in 1st gear up to certain speed, then in 2nd gear up to some higher speed & so on. Ultimately after vehicle speed reaches max. speed during the cycle which will be in higher gear, vehicle speed is reduced by releasing accelerator pedal.

When vehicle is being accelerated in each gear, there is extra fuel consumption corresponding to increase in kinetic energy in addition to normal consumption due to rolling resistance, air resistance & engine friction. Even though in lower gears fuel consumption is high, extra fuel consumed for acceleration is only for providing increase in kinetic energy irrespective of gear in which it is accelerated.

If this extra fuel would have been used for normal running in respective lower gear, fuel consumption rate would have been much higher & vehicle would have covered less distance.

But energy produced by this extra fuel is stored in the vehicle & used for covering some distance when vehicle slows down in higher gear. In higher gears, fuel consumption or energy consumption for running the vehicle is less. Therefore vehicle is able to cover more distance. Higher the gear, distance covered will be more. e.g. if vehicle slows down in top gear, it will cover more distance than when vehicle slows down in 4th gear.

From all the above points, it can be seen that even though extra fuel is consumed in lower gears, it is utilized at lower rate of fuel consumption in higher gear.

Therefore shifting to higher gear is economical even if we are likely to slow down the vehicle. But since we are not accounting for extra fuel consumed for acceleration & we are already accounting fuel even while slowing down depending upon vehicle speed & gear, there is no need for separate accounting of extra fuel for acceleration.

In City traffic signals are close by & therefore there is natural tendency of not shifting to higher gear even though it is possible to shift to higher gear, thinking that even if gear is shifted to higher gear, vehicle speed has to be reduced thereafter. But in 7.5 above we have seen that shifting to higher gear is economical, therefore we should shift to higher gear in this case also.

---oOo---

8. Moving On a Gradient

We all know that for the vehicle to climb up a hill needs extra quantity of fuel. This is due to increase in Potential Energy of the vehicle as it moves upward. To understand the quantity of extra fuel, we need to first understand amount of energy required to move the vehicle upward through a certain height.

Energy required to move an object upward through vertical height 'h' is given by the formulae –

Energy for moving up the gradient –

E
$= mgh$
$= W \times h$
where h – height in meters
W- weight of vehicle in kg
m – mass of vehicle = W/g
g – gravitational acceleration
$=9.81 \ m/sec^2$

Let us consider a vehicle moving an upward gradient of 1 in 'Y'. In this case when vehicle moves on the slope through a distance of Y meters, it moves upward through height of 1 meter. Therefore when vehicle moves through a distance of 1 km i.e. 1000 meters it will move upward through height of 1000/Y meter.

Therefore Energy/km is –

E

$= W \times h$
$= W \times 1000/Y \quad$ in kg-m / km
$= W / Y \quad$ in kg-km / km   …   (8.1)

We know that FC = Cf x E, Therefore, fuel consumption for moving up the gradient –

FCg

= Cf x E

= Cf x W / Y … (8.2)

For our model car, as given in Table 5.1, at speed of 43 kms/hr, in 3rd gear, fuel consumption is 40 cc/km & KMPL is 25.

If the vehicle starts moving upward on a gradient, then there will be extra fuel consumption as given in equation (8.2). Table 8,1 shows additional FCg for moving on different gradients having slope of 1 in Y.

Table 8.1

Fuel Consumption for moving up the gradient With slope of 1 in Y

Y	FCg	FC	FC total	% Increase	KMPL
50	26	40	66	64 %	15
40	32	40	72	80 %	14
30	43	40	83	107 %	12
20	64	40	104	160 %	10

FCg – extra fuel consumption
due to gradient

From the above table we can see that for moving upward on a gradient with a slope of 1 in 20 meters, there is an increase in fuel consumption by 160 % & KMPL reduces from 25 KMPL to 10 KMPL. Even on a moderate slope of 1 in 50, there is an increase in fuel consumption by 64 %.

Even though extra quantity of fuel is consumed to move up the gradient, there is no energy consumption. This extra fuel is used to increase the Potential Energy of the vehicle. This energy is stored in the vehicle & gets released when vehicle start coming down the hill & is utilized to get extra distance without addition of fuel. This is similar to release of K.E. as we have seen in 'Acceleration'.

Many times this release of P.E. even results in acceleration of the vehicle i.e. P.E. gets converted to K.E. But ultimately this K.E. is also used to cover extra distance without consuming any fuel.

Therefore it is not necessary to account for extra fuel needed to move up on the gradient for estimating KMPL for round trip i.e. when we are starting from destination 'A' & ultimately returning back to 'A'.

However if any one going to hill station is specifically interested to know one way KMPL while moving upward separately & moving down separately then it will be necessary to account for extra fuel in upward trip to hill station & reduction in fuel consumption in down ward trip due to release of P. E. This can be done separately as follows –

Energy required for moving the vehicle upward by height 'h' is given by –

E

$= mgh$

$= W \times h$ in kg-m

$= (W \times h) / 1000$ in kg-km

Therefore –

Fuel consumption
$= C_f \times E$
$= C_f \times W \times h / 1000$ cc

Let us say height 'h' of hill station with reference to your starting point is 500 meters & W = 1600 as in case of our model vehicle , then –

Fuel Consumed

= 0.8 x 1600 x 500 /1000 cc

= 640 cc
= 0.640 liter

i.e. when we are moving upward to the hill station we will be consuming 0.642 liter extra for increasing Potential Energy of the vehicle, but when we come down hill there will be release of this P.E., therefore we will need 0.642 liter less.

Even though it is true that there is no need to account for extra fuel used for moving up the gradient, KMPL in hilly region will be low.

This is basically due to operation in lower gears both while climbing up the hill & also while coming down. This effect of operation in lower gears we have already seen in the chapter on 'Engine Friction'.

Also use of brake for keeping vehicle speed under control is likely to be more while coming down the slope. This also results in increase in fuel consumption.

Even though there can be increase in fuel consumption due to other reasons as seen above, the fact remains that there is no need to account for extra fuel consumed for moving up the gradient. Losses due to operation in lower gears we are already accounting for & losses due to braking we will study in next chapter.

Further, unless gradient is very steep & for much larger distance, many times there is no need to shift to lower gear or any necessity of use of brakes, particularly, on city flyovers & gradients other than those in hilly regions.

Important point to remember is that we should operate in highest possible gear & minimize use of brakes.

Summary –

1. Extra fuel is consumed for moving up a gradient & correspondingly there is drop in KMPL. But there is no energy consumption. This energy is stored in the vehicle in the form of 'Potential Energy'.

2. When vehicle comes down the gradient, this 'Potential Energy' is released & vehicle covers extra distance using this energy.

3. There is no need to account for extra fuel consumed for moving up a gradient as it is compensated while coming down.

But in hilly region there can be increase in consumption due to operation in lower gear both while moving up & moving down & also use of brakes for controlling speed of the vehicle while coming down.

---oOo---

9. Braking Losses

In the chapter on 'Acceleration' we have seen that there is extra fuel consumed for acceleration, over & above that required for steady speed running of the vehicle. But there is no energy consumption. This extra energy produced is stored in the vehicle in the form of kinetic energy & gets released when vehicle speed is reduced & we are able to cover extra distance without consuming fuel. Therefore we have stated that there is no need to account for extra fuel consumed for acceleration.

But when we are applying brake for stopping the vehicle or for reducing speed, there is loss of part of this kinetic energy & vehicle covers less distance. Therefore this loss needs to be accounted for.

When accelerator pedal is released & vehicle is allowed to slow down either without braking or with braking, there is no fuel consumption & vehicle is running only using kinetic energy of the vehicle.

9.1 Release accelerator pedal early before stop to reduce braking losses – Simple way of reducing braking losses is to anticipate situation ahead & release accelerator pedal early to allow vehicle to slow down & then apply brake when speed is considerably reduced. By doing this, we are able to cover maximum distance using kinetic energy of the vehicle & without consuming any fuel.

But generally there is a tendency to keep on accelerating or maintaining constant speed till we approach very close to stop & then apply brake to stop vehicle in short distance. In this case, only small distance is covered using kinetic energy of the vehicle & remaining energy is wasted in braking.

We have to remember that this kinetic energy is gained by vehicle consuming extra fuel during acceleration.

Let us understand the difference in the following two cases for the vehicle running at some speed & consuming fuel at the rate of 50 cc/km –

Case I – Accelerator pedal is released 700 meters before stop – In this case, there is no fuel consumption for last 700 meters before stop. Prior to 700 meters it was consuming at the rate of 50 cc/km

Case II – Accelerator pedal is released 200 meters before stop – In this case, there is no fuel consumption for last 200 meters. But prior to 200 meters it was consuming at the rate of 50 cc/km.

Now, let us see the difference in fuel consumption in these two cases –

In Case I, last 700 meters are run without fuel. But in Case II, only last 200 meters are run without fuel. It means that in Case II, fuel has been consumed for additional distance of (700 – 200) = 500 meters as compared to Case I at the rate of normal consumption of 50 cc/km.

From the above it will be clear that fuel saved is same as fuel required for running the distance saved at normal rate of fuel consumption.

9.2 - % increase in fuel consumption due to braking losses – In chapter on acceleration we have seen extra fuel required for providing kinetic energy (K.E.) of vehicle at different vehicle speeds (Pl. refer Table 7.1).

Table 9.1 shows fuel consumed in cc for providing kinetic energy of the vehicle at different vehicle speeds. (values taken from Table 7.1)

If average fuel consumption of the vehicle is say 50 cc/km, then we can calculate how much distance vehicle will be able to run using this stored energy. e.g. at 60 kms/hr, fuel for K.E. is 18 & fuel consumption is 50 cc/km. Therefore with 18 cc fuel vehicle can cover a distance of (18/50) x

1000 meters. This calculated distance is shown as distance without braking.

But in actual practice, due to braking vehicle is stopped in less distance. Difference between distance that can be covered without braking & with braking is shown as distance lost.

In this Table speed range of 100 -60 -0 indicate that accelerator pedal is released & vehicle is allowed to slow down from 100 to 60 kms/hr & then brakes are applied.

From the Table we can see that at 100 kms/hr K.E. is sufficient to cover a distance of 1000 meters without braking & at 60 kms/hr K.E. is sufficient to cover a distance of 360 meters. It means that when we are allowing speed to reduce from 100 kms/hr to 60 kms/hr vehicle will be able to cover a distance of 1000 -360 =640 meters. Further, when speed is reduced from 60 kms/hr to 0 with braking, further distance of 70 meters is covered as given in table. i.e. total distance of 640 + 70 =710 meters are covered when we are going in 100-60-0 type braking.

Table 9.1

V in Kms/hr	Fuel in cc for K.E.	Distance Without braking	Distance With braking	Distance Lost in Meters 'd'
100	50	1000	200	800
80	32	640	120	520
60	18	360	70	290
50	13	260	50	210
40	8	160	30	130
100-60-0	50	1000	710	290
60-40-0	18	360	230	130

Distance lost 'd' shown above is for one application. If 'n' is no. of brake applications/km then distance lost / km – 'D' will be n x d. It means that 'D' meters are lost in every 1 km or 1000 meters. From this we can calculate % of distance lost. Since fuel is already consumed for this distance lost, this % distance lost also represents % increase in fuel consumption due to braking.

Accordingly, distance lost /km – D & % increase in fuel consumption for different speed ranges & various values of 'n' are calculated & shown in Table 9.2.

Table 9.2

V In Kms/hr	Distance Lost 'd'	'n'	Distance Lost per Km 'D'	% Increase In FC
Highway -				
100-0	800	0.1	80	8 %
100-0	800	0.2	160	**16 %**
100-60-0	290	0.1	29	3 %
100-60-0	290	0.2	58	6 %
City -				
60-0	290	1	290	29 %
60-0	290	1.5	435	**44 %**
60-40-0	130	1	130	13 %
60-40-0	130	1.5	195	20 %

Where d – distance lost in meters/ brake application
n – no. of brake applications/km
& D – distance lost in meters/km

Following observations can be made from % increase in fuel consumption due to braking shown in Table 9.2 –

A. Highway Running -

In highway running no. of brake applications are low, but If we start applying brake when vehicle speed is high, then % increase in fuel consumption is high. e.g. -

For one brake application in 5 kms (n = 0.2), if we start applying brake when speed is 100 kms/hr then % increase in fuel consumption is 16 %. But instead if we allow the vehicle to slow down to 60 kms/hr & then apply brake then % increase in FC will reduce to 6 %.

B. <u>City Operation</u> –

In city operation vehicle speed is low & correspondingly, distance lost / brake application is also low. But no. of brake applications are more. Therefore, % increase in FC is much higher. e.g. –

If we start applying brake when speed is 60 kms/hr, % increase in FC is as high as <u>44 %</u>. But if we allow speed to reduce to 40 kms/hr & then apply brake then % increase in FC will be below 20 %.

Summary –

1. To reduce braking losses, it is necessary to anticipate situation ahead, release accelerator pedal early to gainfully utilize kinetic energy of the vehicle & then apply brake when vehicle speed is considerably reduced. If brakes are applied when vehicle speed is high, then braking losses are many folds higher.

2. We should avoid driving habits which results in unnecessary braking.

---oOo---

10. Idling Losses

When vehicle is standing at a signal, engine is running at idling r.p.m. Under this condition, for our model vehicle there will be fuel consumption as given in Table-4.1 i.e. 0.29 liters / hr.

In highway running % of time spent in idling to the total time for the trip is negligible & will be even less than 2 %. But in City operation % of time spent in idling can be 25 % or even more.

In case of idling, since vehicle is stationary & there is no distance covered, we cannot calculate fuel consumption in cc/km or KMPL.

Therefore to calculate the effect of idling consumption, on trip KMPL we will have to adapt different method as shown below –

Let us take an example, of our model car –

In this case, idling fuel consumption is 0.29 litres/hr or 290 cc/hr. Let us assume that % of idling time to total time for the trip is say 25 %. Then idling fuel consumption for the trip will be 25 % of 290 cc /hr i.e. 72.5 cc/hr.

If the average vehicle speed is say 30 kms/hr, then this idling consumption of 72.5 cc in 1 hour gets divided over 30 kms.

Therefore,-
Fuel consumption due to idling
= 72.5 / 30
= 2. 41 cc/km
If average kmpl for the trip is 18, then –

FC

= 1000 / KMPL

= 1000 /18

= 55.56 cc/km

Now, in average fuel consumption of 55.56 cc/km, there is an increase of 2.41 cc/km.

Therefore –

% increase in fuel consumption due to idling

= 4.34 %.

From the above example, it can be seen that even for 25 % idling time, net increase in fuel consumption is 4.34 %.

But we also need to consider following points –

When engine is running at idling speed we are consuming fuel at the rate of 290 cc/hr i.e. 4.83 cc/minute.

If we stop engine then –

i. for starting engine, it requires about 3 cc. of fuel.

ii. Also for starting the engine we have to crank the engine using electrical starter for which we are consuming battery power. This loss of battery power is then replenished during charging of battery for which extra fuel is required.

iii. It is observed that when we are stopping at the signal, but at some distance away from the signal, many a times, scooters & 3 wheelers will find some narrow gaps & will move forward leaving space for other vehicles to move forward. In that case, we will also have to move forward. I had many a times observed vehicles moving forward 2 to 3 times at some of the signals. If we keep on cranking the engine several times like this in short time interval, then we will be facing the problem of battery weak & we will getting stranded on the road.

Considering all the above points, it is economical to stop the engine at signals if the time is likely to be more than 2 minutes.

Therefore even though total loss due to idling is 4.54 %, net saving what we can achieve by stopping the engine at signals/stops is less than 2 %. This saving is much smaller as compared to saving by taking care of other factors. But many times, this point is found to be overemphasized to such an extent as if it is the only major point for saving fuel.

No doubt, even 1 or 2 % saving is of important to us, but one should do it only when waiting time at signal <u>at the same place</u> is likely to be more than 2 minutes.

---oOo---

11. Driving For Better Fuel Average

We have seen fuel consumption for constant speed running of the vehicle due to rolling resistance, air resistance & engine friction for various speeds & gears. Average of all these values depending upon operating condition gives us average fuel consumption for the trip.

We have also seen that there is no need to account for extra fuel consumed for acceleration or moving up a gradient for accounting trip fuel consumption.

Therefore for knowing total fuel consumption, we need to add, % increase in consumption due to braking losses & idling losses only.

11.1 Driving in City Traffic –

In city traffic many a times we have to operate at very low speeds & in lower gears. Also no. of brake applications in City operations are very high as compared to highway running.

11.1.A – Try to run the vehicle in the most economical zone –

Always try to run the vehicle in most economical zone. Normally it is around lowest speed in top gear.

At lower speeds & in lower gears fuel consumption can be more by 100 % or even higher. Therefore we should try to achieve speed range of most economical zone as early as possible.

11.1.B – Always drive in highest possible gear –

i. At any vehicle speed, higher gear gives better KMPL -

In City traffic, normally vehicle speed is decided by traffic conditions on the road, but we have the option of selecting proper gear. If we select higher gear, fuel consumption is lower. But if we operate in lower gear, increase in fuel consumption can be 26 % to 90 %. (Pl. refer Table 5.3) Particularly, at lower speeds, operating in just one gear lower, % increase in fuel consumption is more than 50 %.

ii. Higher gear is economical even while slowing down the vehicle.

As seen in the chapter on acceleration even while slowing down the vehicle if the vehicle is in higher gear, energy requirement of the vehicle is low & vehicle can cover more distance using kinetic energy of the vehicle. This results in saving of fuel.

iii. If faster acceleration is not possible due to traffic conditions then shift to higher gear early –

Normally, we will be able to shift to higher gear, when engine r.p.m. is more than around 2600 r.p.m. instead of continuing in lower gear till max. r.p.m. of 4000 is reached.

Operating in just one gear lower, particularly at higher r.p.m.s results in increase in fuel consumption by 50 % or more.

11.1.C. Faster Acceleration is economical in lower gears –

We have seen in the chapter on 'Acceleration' that whether faster acceleration or slower acceleration has no effect on trip KMPL.

But when we accelerate at faster rate, we will be able to shift to higher gear early & thereby reduce % of operation in lower gears & avoid unnecessary increase in fuel consumption.

11.1.D Release Accelerator Pedal early before Stop –

We have seen that when we slow down the vehicle, by releasing accelerator pedal, there is no fuel consumption & vehicle is run using kinetic energy of the vehicle.

This kinetic energy is gained at the cost of extra fuel consumed during acceleration.

Therefore we should anticipate the situation ahead & release accelerator pedal considerable distance before stop to make maximum use of kinetic energy.

If we continue accelerating the vehicle or running the vehicle at constant speed till we come close to the stop, then vehicle continues to consume fuel at the normal rate of consumption. Fuel consumption stops only when accelerator pedal is released.

11.1.E. - Slow down in Neutral where ever possible –

Referring to Graph 5.4, we have seen that fuel consumption due to engine friction particularly in lower gears is very high in the range of 50 % to 80 % & even at higher speeds & higher gears it is in the range of 20 % to 50 %.

Therefore while slowing down the vehicle, when we do not need any engine power we can avoid this huge loss by shifting to neutral. Since loss due to engine friction is 50 %, by shifting to neutral we can cover double distance using kinetic energy of the vehicle without any fuel.

Many surveys taken about City operation reveal that nearly 30 % to 40 % of time, vehicle is slowing down or run with accelerator pedal released. Considering both the above points we can see that there is a scope of saving nearly 15 to 20 % (0.5 x 0.3 = 0.15) of fuel.

But for getting benefit of extra distance without fuel, we must anticipate situation ahead & release accelerator pedal early correspondingly. Otherwise whatever energy saved will be wasted in braking.

However, this method needs to be used with due care as given in chapter 5 i.e. difficulty in re-engaging gear & risk involved while going down steep gradient.

11.1.F. - Idling – Stop engine, if waiting time at signal or stop is likely to be more than 2 minutes – In City operation % of idling time to total time can be more than 25 %. This can increase fuel consumption by about 5 %.

11.1.G. - Driving on City Flyovers – During driving training, it is taught that when we are climbing up the gradient we should use lower gears & it is also taught that whatever gear is required to be used for climbing the hill, same gear should be used for going down the same gradient.

However, on most of the City flyovers where there is free flow of traffic, it is observed that particularly in case of cars we can easily pass over the flyover in top gear if we are already at speed above 60 kms/hr. Normally, city flyovers do not have steep gradient & most of the cars are able to climb such gradients in top gear & can have reasonable level of acceleration also. In such case, it is better to continue in top gear only & get better kmpl instead of unnecessarily lowering down the gear causing increase in fuel consumption.

11.1.I - For vehicles with auto transmission, when vehicle stops, shift gear position to neutral – If the vehicle is with auto transmission, ensure that gear position is shifted to neutral position when vehicle stops.

In case of vehicles with manual transmission, when vehicle speed is reduced, engine speed is also reduced & when engine speed drops to idling speed we have to necessarily shift gear position to neutral as otherwise engine will start giving jerks & will ultimately stall.

But in case of vehicles with auto transmission it is possible to keep the gear engaged. In this case torque converter slips & allows the engine to run. But under this condition there is load on the engine & correspondingly consumes much higher fuel than our normal consumption at idling.

Therefore it is necessary to ensure that gear position is shifted to neutral when we are stopping the vehicle with auto transmission, but be careful to prevent roll back of the vehicle. In gear position it was prevented, but in neutral it will not be prevented.

11.2 - Driving on Highways –

In highway running, normally speed is high. No. of stops or signals are comparatively very low & correspondingly no. of brake applications/km are low. But due to high speed operation, braking loss/brake application can be high. Therefore one needs to be cautious in controlling unnecessary braking losses.

Also even in intercity trip, most of the time vehicle has to move through City traffic for a distance of more than 20 to 30 kms at both ends. Therefore all precautions, given for City operations, needs to be taken care of.

11.2.A - Avoid very high speeds – Air resistance is proportional to V^2. Therefore fuel consumption increases sharply with increase in vehicle speed. We had seen that % increase in fuel consumption at higher speeds, compared to fuel consumption at 60 kms/hr in top gear, can be 100 % or 200 % or even more. Therefore restrict your speed. After all choice is yours to decide how much extra fuel cost you are ready to bear – 100 % or 200 %. [at 100 kms/hr, % increase is 83 %]

11.2.B –Use of Overdrive –In olden days, gear ratio of 1: 1 was being called as 'Top gear' & upper gears were being called as overdrive. In case there are two overdrives, then they were being called as overdrive I & overdrive II.

If you are operating the vehicle at a speed more than min. speed required for use of overdrive I or II, then that overdrive should be used.

However rate of acceleration available in overdrive will be less than that in top gear. Therefore, we may choose to shift to overdrive,   after attaining desired speed at which we want to drive the vehicle.

11.2.C - Minimize Braking Losses –

i. Release Accelerator Pedal early before Stop – We have seen this in City operation. But in highway operation even though no. of brake applications are less, speed is much higher. Therefore it is necessary to anticipate situation ahead & release accelerator pedal early before stop to utilize kinetic energy of the vehicle to the maximum extent.

ii. Avoid unnecessary braking – Many people are in the habit of attempting to overtake other vehicles without properly assessing the situation ahead & then when they realize that they just cannot overtake the other vehicle, they slow down the vehicle by applying severe brakes.

In such cases, brakes are being applied at high speed & moreover there is no chance to release accelerator pedal early to gainfully utilize kinetic energy of the vehicle. We had seen in chapter on braking that losses due to reducing speed only from 120 kms/hr to 100 kms/hr can be more than 14 %. These type of brake applications are totally avoidable.

---oOo---

12. Scope for Design Improvements

We have seen various factors that affect fuel consumption of the vehicle. In the last chapter we have seen what precautions that we can take to get best possible fuel average during driving. Now, we will see what design improvements can be done for better fuel economy.

12.1 - <u>Reducing Air Resistance</u> – We have already studied various factors that affect air resistance. We have seen that air resistance increases in proportion to square of vehicle speed. Further modern trend is to go for the vehicles with increased max. speed. Even with speed level of 100 to 120 kms/hr, we have seen that nearly 50 % to 60 % of fuel consumption is on account of air resistance only. With increase in speed this % will still go higher. Therefore it has now become absolutely essential for designers to take all out efforts, to reduce air resistance.

12.1.A - <u>Reducing Frontal Area of the Vehicle</u> – Even though it is called frontal area, it is the total area which obstructs the flow of air. Critical examination will have to be done to find out points where there is scope for reducing frontal area. For example -

1. One can think of reducing the width of a car or bus body above shoulder level of the sitting passenger.

2. For reducing width of a car one can reduce no. of passengers in one row & increase no. of rows.

3. For reducing height of a car already many models have altered sitting position. i.e instead of sitting with legs in upright position now you have to sit with legs in stretched condition.

4. Racing cycles or motorcycles are deliberately designed in such a way that rider has to considerably lean forward to access the handle. This is because when rider sits in upright

position, frontal area is more. But when the rider leans forward frontal area is drastically reduced.

We know that nearly 40 to 50 % of fuel consumption is on account of air resistance in the speed range of 80 to 100 kms/hr. <u>Therefore 10 % reduction in frontal area will reduce air resistance by 10 % & fuel consumption will reduce by 4 % to 5 %</u>.

12.1.B - <u>Reducing co-efficient of air resistance – Cd</u> - We know that if front face of the vehicle is kept vertical, air resistance is at maximum & as we go on tilting front face backward co-efficient of air resistance goes on reducing. Similarly giving proper slope at the rear is also equally important. Similarly, we can also provide curved portion in the front so as to divert the air sideways. All these points are applicable for rear also.

In case of cars, considerable improvements have been done & co-efficient of air resistance - Cd has been brought down from the level of more than 0.7 in the olden days to almost 0.4. But considering the way max. speed of vehicles is increasing, it needs further improvement.

In bus & truck segments there is practically no improvement in Cd. Even today most of the buses & trucks have Cd in the range of 0.75 to 0.8. There is lot of scope for improvement in this segment.

1. <u>In case of buses</u> –

i. Front face is still almost vertical, backward tilt angle is hardly 15 degrees. One can definitely provide more backward tilt very easily.

ii. Modern trend is to use full height windscreen glass. But providing glass above the range of driver's vision is of no use. Further this, results in unnecessary tremendous increase in cost. Instead we can provide windscreen glass sufficient to cover required range of driver's vision & the upper portion can be provided with sufficient slope to reduce air resistance as shown in dotted lines in Fig. 12.1.

iii. Also for rear portion of bus it is found that absolutely no consideration is given for reducing air resistance. It is possible to provide some slope & curvature at the rear end of the bus.

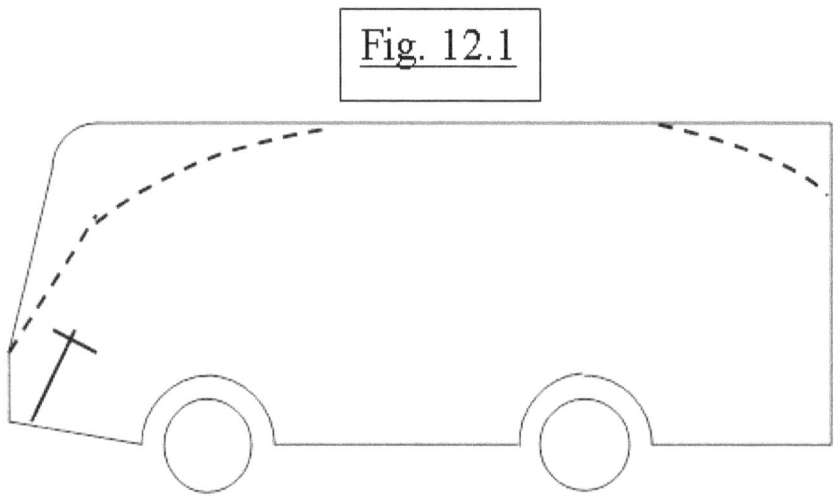

Fig. 12.1

By stream lining the body as above it is possible to reduce co-efficient of air drag from 0.8 to 0.5. It means a reduction of 37.5 % in air resistance.

It can be seen from Table A.1 given in Annexure for buses that % of fuel consumption due to air resistance in total consumption for the speed of 80 kms/hr is around 45 %.

Therefore by changing the shape of body as above, there will be reduction in fuel consumption by 37.5 % of 45 % i.e. by nearly 17 %.

2. In case of medium & heavy duty trucks it is observed that as compared to driver's cabin, goods compartment behind, is of much greater height & front face of the goods compartment is kept vertical. As we know, for such vertical

surface, co-efficient of air resistance Cd is about one. But if we provide an additional panel as shown in the Fig. 12.2 below we can reduce Cd to the level of nearly 0.4 for this area. This means reduction in Cd by 60 %.

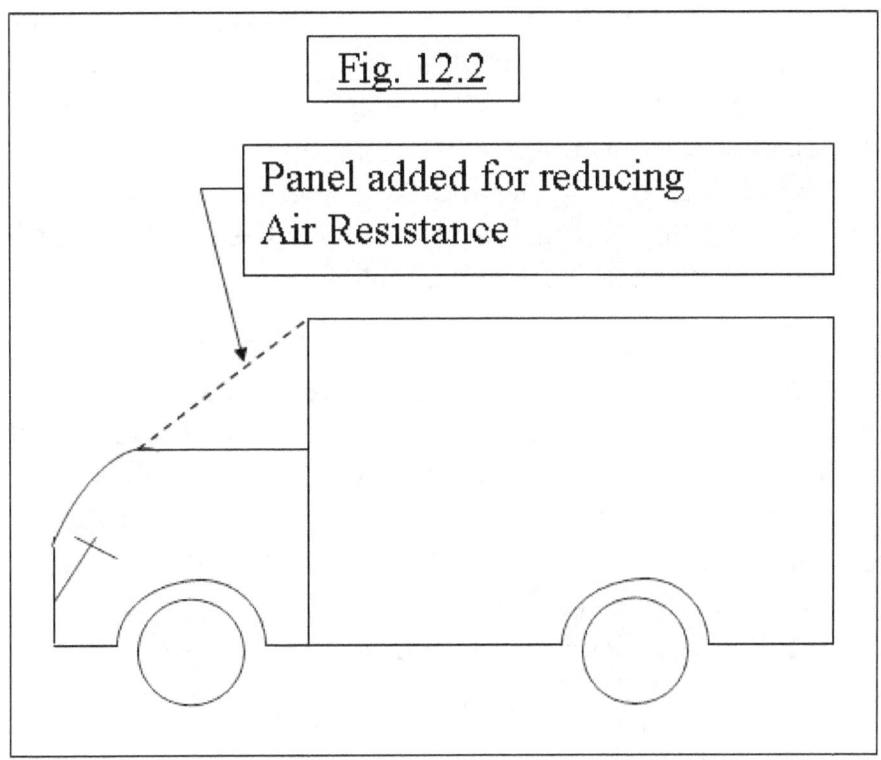

If this portion is say about 40 % of the total area. Then Cd for the total frontal area will reduce by 60 % of 40 % i.e by nearly 24 %.

Further since fuel consumption due to air drag is about 40 % of the total consumption at 80 kms/hr, therefore net reduction in fuel consumption will be 40 % of 24 % i.e. 10 %.

This simple panel added reduces fuel consumption to the tune of 10 % due to reduction in air resistance.

Same method can be used for luggage carriers at the top of cars or buses. But in that case, it is better to keep panel adjustable so that it can be used only when there is a luggage on the top & can be adjusted to suit height of luggage.

3. <u>In case of scooters & motor cycles</u> – In case racing cycles & motor cycles handle position is so designed that rider has to lean forward. This reduces air resistance considerably. Even though incorporating this feature as it is may not be acceptable to non racing people. But one can explore the possibility of having some compromise in the lean forward position which will be acceptable to the people.

4. <u>Engine Compartment</u> – In total frontal area engine compartment area is comparatively smaller about 20 % to 30 %. But in view of, increased maximum speed of the vehicles it has become essential to take every effort to reduce air drag. In many cases particularly in tempos, trucks & buses it is found that panels at back of the engine compartment are not properly designed for smooth air flow & there is considerable scope for improvement.

Reducing air drag of the engine compartment can give us some reduction in the fuel consumption even though small but at the same time it will improve efficiency of cooling system.

12.2 - Rolling Resistance – For radial tyres, co-efficient of rolling resistance is nearly 10 % lower as compared to cross ply tyres. Therefore using radial tyres in place of cross ply tyres gives 10 % reduction in rolling resistance. Rolling resistance is nearly 20 to 30 % in case of our model vehicle for speed of 40 to 60 kms/hr. as can be seen in the chart given earlier & at higher speeds as well as in lower gears it is much less.

Therefore we can get reduction in fuel consumption by nearly 2 to 3 % (i.e. 10 % of 20 to 30 %) by use of radial tyres.

12.3 - Engine Friction – Fuel consumption on account of engine friction forms a major portion of total fuel consumption. From the data we have already worked out, it can be seen that in top gear, fuel consumption due to engine friction is 15 % to 25 % of total fuel consumption & in City operations where maximum speed is restricted to 50 to 60 kms/hr, it is 50 % to 80 %. (Refer Graph 5.4)

Since major portion of our fuel consumption is on account of engine friction alone, we must pay special attention to it.

We have already discussed about reducing the effect of engine friction by shifting to higher gear, slowing down in neutral etc. in the chapter on 'driving'.

We will now see what the designers can do to reduce fuel consumption on account of engine friction.

Use of Freewheel – We have already seen in the chapter 11, that when we do not need any power from engine, we can just shift to neutral to avoid losses due to engine friction. But we had some difficulties i.e. if it becomes necessary to re-engage gear, then synchronizing was difficult. This difficulty can be easily overcome, if freewheel is provided between engine & wheels.

Freewheel is a one way clutch. It transmits torque only in one direction i.e. from input to output side. But when output shaft starts rotating faster than input shaft, freewheel slips & do not drive input shaft. This is basically same as bicycle freewheel. In case of bicycle, when we pedal bicycle, wheel rotates. But when wheel is rotating pedal does not rotate.

Further in case of freewheel, input shaft gets connected to output shaft only when speed of input shaft is equal to speed of output shaft. Therefore there is no problem of synchronizing as in case of re-engaging gears.

Various surveys taken about City operation indicate that nearly 40 % of time, vehicles are run with accelerator pedal in released condition.

We have already seen that fuel consumption due to engine friction is 50 % to 80 % of total fuel consumption. Therefore by providing freewheel there will be reduction in fuel consumption to the tune of 50 to 80 % for nearly 40 % of the total time of operation. It means that total saving will be 20 % to 32 %.

Obviously, one has to take care of following points while providing freewheel –

1. Arrangement for locking freewheel or alternate arrangement for going down the steep gradient.

2. In order to facilitate engagement of reverse gear, either freewheel will have to be located between engine & gearbox or alternate arrangement of locking freewheel or some other arrangement will have to be provided.

In case of motorcycles, where there is no reverse gear, there will be no such problem.

Further, once this method is developed, one can accelerate at faster rate say from 60 kms/hr to 80 kms/hr & then allow the vehicle to slow down to 60 kms/hr & then again accelerate faster to 80 kms/hr. By using this method one can considerably increase % of time vehicle is run with absolutely no engine friction. In lower gears, this % of time will be very high.

As seen above by providing freewheel, we can considerably reduce fuel consumption even in city traffic by more than 30 %. Also saving of more than 10 % can be achieved in highway running.

Some people think that in absence of engine as brake, it will lead to some problem. However, it is a fact that force due to engine friction at the wheels, is less when vehicle is in high speed in top gear. But when vehicle is at slow speed & hence in lower gears, this force is multiplied as per gear ratio which is practically acting as brake. In short, due to engine friction there is no much of force when it is required at high speeds, but at low speeds when opposing force is not required, this force is much higher.

In any case, we cannot ignore introduction of freewheel considering huge potential for saving in fuel.

---oOo---

APPENDIX (Model Bus)

Specifications of Model Bus

W	10000 kg	Gear ratios	
A	6.37 sq. m.	1st	4.57
Engine	6000 cc	2nd	2.76
Tyre	10 x 20	3rd	1.66
Diff. ratio	5.88	Top	1
Co-efficients			
Crr	0.01		
Cd	0.8		
Cf	0.8		

Table A.1
FC in cc/km & KMPL (**BUS**)

r.p.m	1200	1500	1800	2100	2400
FC$_{NL}$	2.08	2.9	3.84	4.9	6.08

Top -					
V	40	50	60	70	80
FCef	52	58	64	70	76
FCrr	80	80	80	80	80
Fcar	32	49	71	96	126
FC	164	187	215	246	282
KMPL	**6.1**	**5.3**	**4.7**	**4.1**	**3.5**

3rd -					
V	24	30	36	42	48
FCef	86	96	106	116	126
FCrr	80	80	80	80	80
Fcar	11	18	26	35	46
FC	178	194	212	231	252
KMPL	**5.6**	**5.2**	**4.7**	**4.3**	**4.0**

2nd -					
V	15	18	22	25	29
FCef	143	160	176	193	209
FCrr	80	80	80	80	80
Fcar	4	6	9	13	17
FC	227	246	266	286	306
KMPL	**4.4**	**4.1**	**3.8**	**3.5**	**3.3**

1st -					
V	9	11	13	15	18
FCef	238	265	292	320	347
FCrr	80	80	80	80	80
FCar	2	2	3	5	6
FC	319	347	376	405	433
KMPL	**3.1**	**2.9**	**2.7**	**2.5**	**2.3**

Graph A.1
KMPL (BUS)

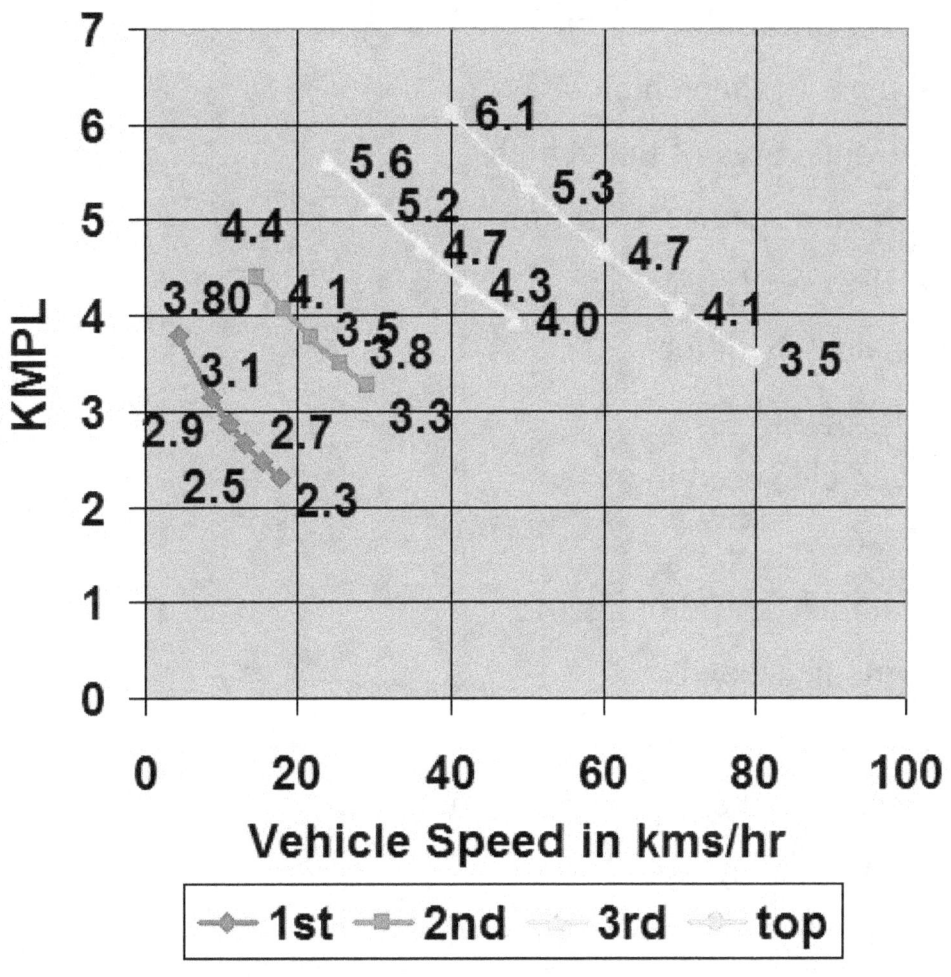

Vehicle Speed in kms/hr

Legend: 1st — 2nd — 3rd — top

Abbrevations

KMPL – Kilometers Per Liter

MPG – **M**iles **P**er **G**allon

W – weight of the vehicle

m – mass of the vehicle = W/g

A – frontal area of the vehicle

g - gravitational acceleration

= 9.81 m/sec2

rho – density of air

= 1.14 kg/m3

v – vehicle speed in m/sec

= V/3.6

V – vehicle speed in kms/hr

t – time in seconds

h – height in meters

n – no. of brake applications/km

F – Force in kg.

Frr – Force due to rolling resistance

Far – Force due to air resistance

Fef – Force due to engine friction

E – Energy in kg-km/km

FC – fuel consumption in cc/km

FCrr – fuel consumption due to rolling resistance

FCar – fuel consumption due to air resistance

FCef – fuel consumption due to engine friction

Crr – co-efficient of rolling resistance

Cd – co-efficient of air resistance

Cf – conversion factor

---ooo---

www.ingramcontent.com/pod-product-compliance
Lightning Source LLC
Chambersburg PA
CBHW081311170526
45166CB00011B/3480